IGCSE
Study Guide

for Physics

Mike Folland

HODDER EDUCATION
AN HACHETTE UK COMPANY

Hachette UK's policy is to use papers that are natural, renewable and recyclable products and made from wood grown in sustainable forests. The logging and manufacturing processes are expected to conform to the environmental regulations of the country of origin.

Orders: please contact Bookprint Ltd, 130 Milton Park, Abingdon, Oxon OX14 4SB. Telephone: (44) 01235 827720. Fax: (44) 01235 400454. Lines are open 9.00–5.00, Monday to Saturday, with a 24-hour message answering service. Visit our website at www.hoddereducation.co.uk.

© Mike Folland 2005
First published in 2005
by Hodder Education,
an Hachette UK company
338 Euston Road
London NW1 3BH

Impression number 10 9 8
Year   2012 2011

Cover photo TEK Image/Science Photo Library
Illustrations by Mike Humphries
Typeset in Bembo 12/14pt by Pantek Arts Ltd, Maidstone, Kent
Printed and bound by CPI Group (UK) Ltd, Croydon, CR0 4YY

A catalogue record for this title is available from the British Library

ISBN: 978 0 7195 7903 5

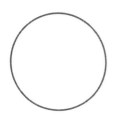

Contents

Contents

Certain sections of this book are shaded . This indicates Supplement material and should be studied by students following the Extended curriculum.

Introduction

Structure of the book

This text has primarily been written to support students in the study of Physics to IGCSE. The five topics covered in this book correspond to the five topics in the IGCSE examination syllabus.

The syllabus has two components, the Core and the Supplement; the latter defines the Extended curriculum. The Core curriculum is graded C to G and the Extended curriculum can be graded A* to G. To differentiate between these components, sections of this book covering the Supplement are shaded. Wherever possible, this practice has been extended to the questions, some of which may involve material from both the Core and Supplement.

Each topic starts with a list of **Key objectives** which specifies the skills and knowledge you will need to have acquired during your study of this topic. The list can also serve as a checklist of your progress in this topic. Each topic also includes a number of essential terms which are defined in a table of **Key definitions**. The main sections of each topic are the **Key ideas**, and it is essential that you both learn and understand these concepts. This is the Physics that you will need for the IGCSE examination. **Common misconceptions and errors** are listed to help you avoid the common mistakes made by students (see page viii).

In each topic there are questions. **Sample questions** are in the style of IGCSE questions. These are followed by an answer given by an imaginary student to illustrate how answers of different quality would be awarded marks. This is then followed by examiner's comments and the correct answer. There are many **Try this** questions to check you have learnt the Physics. You should use these for practice and to assess your understanding and recall of the topic. You will find the answers at the back of the book.

Preparing for the examination is primarily directed at those taking the external IGCSE examination, but much of the advice offered is also relevant to internal examinations.

How to use this book

Who is this book for?

The purpose of this book is to aid any IGCSE Physics student to achieve the best possible grade in the exam. Ideally, you will have good teaching and access to the textbook *IGCSE Physics*, written specifically for this course by Tom Duncan and Heather Kennett. This Study Guide will give you support during the course as well as advice on revision and preparation for the exam itself. However, not all students are in such a fortunate position. For those with only a little qualified teaching and no textbook, this Study Guide provides all the essentials you will need to prepare yourself for the exam.

This book is designed to help teachers with the course content by giving key ideas in summarised form, providing example questions with comments from an experienced examiner, as well as numerous tips and misconceptions to avoid.

The book is also designed to help students when they work on their own. The material is presented in a clear, concise form and there are questions for you to try, many with examiner's comments on the expected answers.

When should this book be used?

One main aim of this book is to help students revise and prepare for the exam at the end of the course. However, to get the most from the book and the whole course, it should be used on an ongoing basis throughout the course. The Key ideas, tips and things to avoid are best learnt as early as possible. The examples will be helpful at any stage. The book will be especially helpful to students preparing for end-of-topic tests and internal exams held at various stages during the course by their school or college.

If you want to get the maximum value from this book, it is strongly advised that you attempt to answer all the questions on paper and not in the book. Then you can repeat the exercises at intervals throughout the course.

How do we learn?

We remember very little of what we have only heard. We remember a little more of what we have heard and seen. We only really learn things properly when we actually do things with the material – **active learning**. Not only is this more interesting and motivating for students, but we actually understand, learn and remember things much better this way.

Active learning has always been promoted by good teachers whose lessons are not just 'talk and chalk'. In a practical subject like Physics, there should be demonstrations with the active involvement of students who may take readings, discuss results and suggest further lines of enquiry. Practical work, usually done in small groups, should ideally involve the students in the planning of the investigation. In any case, the students are actively involved in handling the apparatus and taking readings, and they usually go on to make calculations, draw graphs and deduce conclusions.

Study on your own should also be organised so that you are active. Simply reading through a textbook (or even this Study Guide!) and hoping you will learn just does not work. The **Preparing for the examination** section on pages 96–100 will give advice in detail on how to learn and revise for tests and exams.

Before you start to work with this book, you should ideally have a separate notebook or folder in which to record your work. You might wish to make the occasional comment in your normal exercise book but there will not be space for the amount of extra work that should be done, and some schools may not allow these extra comments. Even though your separate notebook may not be seen by anyone else, you should take the same pride in keeping it as if it were to be marked by the strictest teacher!

As your course proceeds, it will be useful to compare the notes in your exercise book with the topics presented here. There will be differences but this does not mean that either your notes or this book is wrong. In fact, you will learn a lot from carefully comparing the two different versions and you will usually find that the differences are alternative ways of presenting the same information. You should make a note of such points with comments about the differences and similarities.

How to answer different types of questions

● **Calculations**

● **Always show your working.** The first reason for this is that if you write down all your working, you are far more likely to work in the logical way needed to reach correct answers. Also, students often make errors early on in a multi-part question and examiners try hard not to penalise more than once for the same error. If you continue working from an early wrong answer without making a further mistake, you will normally not lose any more marks as long as **the examiner can see clearly what you have done**. Do not expect the examiner to try and guess whether you worked on correctly.

● **Train yourself to set out work logically.** We are not all naturally neat and tidy in how we work but most people can improve with a little effort. Neatness certainly helps logical working and it is important to develop this habit thoughout the course. You may think you can suddenly work logically in the exam but it will not just happen if you have not practised it. Ordinary school or college exercises, and even work from this book which is just for you, should be logically presented with all the steps of your working.

● **Show the units.** Sometimes the question will make it easy for you and give the units. Sometimes units will be asked for. Marks are often given for units, so develop the habit of thinking about the units and writing them down even when not asked for. Scientific quantities are meaningless without units. It makes a lot of difference whether the speed limit on a road is 30 m/s, 30 km/h or 30 mph.

● **Graphs**

● **Drawing graphs.** The axes should be labelled with units. The scales should fill more than half the space available in each direction. Think carefully about whether the origin should or should not be included. Do not use scales with awkward multiples like 3 or 7 – there is no need to fill every last bit of the available space. Points should be plotted carefully to an accuracy of 0.5 mm. This means using a sharp pencil to mark the point with a small cross or a circled dot. If the line of best fit is a straight line, carefully judge its position and draw one thin line with a ruler. If the line is a curve, it should be a single, thin, smooth line through the points. It must not be distorted to pass through every point.

• **Reading off graphs.** Again, you should work to an accuracy of 0.5 mm. Draw vertical and horizontal lines to the axes to show your working.

• Descriptions and logical deduction

• **Logical thinking.** The important thing is not to abandon logic just because words are involved, not numbers. Many questions require step-by-step descriptions and/or deductions. You must be just as logical with words as you would be in working out a question with numbers.

• **How much do I write?** The space available is a rough guide but not a fixed rule. If you need much less space than that provided, think carefully about whether you have missed out something important. If you need a lot more space, you are probably writing about something not in the question or wasting time on extra detail for which the mark scheme will give no more marks.

Examination terms explained

Define	A precise statement is needed
What do you understand by	Give the definition and some additional explanation
State	Give a concise answer; no explanation is needed
Describe	Give clear, positive statements about the situation/objects involved; drawing a diagram is allowed and sometimes very helpful
Explain	You must give reasons and/or underlying theory
Predict	You are not supposed to know the answer from memory but deduce it, usually from information in the question
Suggest	This implies that there is more than one acceptable answer or that candidates are expected to arrive at the answer using their general knowledge of Physics

Common misconceptions and errors

Common misconception

✗ A body in free fall is weightless.
✓ A body in free fall may feel weightless, but weight is the force of the Earth's gravitational field, which still acts on the body. ■

✗ Shows that this statement is wrong
✓ Shows that this is the correct idea
■ Indicates the end of the Common misconceptions and errors section

TOPIC 1 General Physics

Key objectives

- To know the relationships between distance, speed, velocity and acceleration and relate them to moving objects in a variety of situations
- To know that mass is a measure of the amount of matter in an object and that weight is a force related to mass
- To know what is meant by density and how it can be determined
- To know the effects of forces and be able to apply this knowledge to practical situations
- To be able to explain the distinction between scalars and vectors
- To know and be able to explain the relationship between energy, work and power
- To know how pressure is related to force and area

Key definitions

Term	Definition
Speed	$\dfrac{\text{total distance}}{\text{total time}}$
Velocity	Speed in a specified direction
Acceleration	$\dfrac{\text{change of velocity}}{\text{time taken for change}}$
Density	$\dfrac{\text{mass}}{\text{volume}}$
Weight	A force exerted by gravity
Moment	The turning effect of a force
Scalar	A quantity with magnitude (size) only
Vector	A quantity with magnitude (size) and direction
Work	Force × distance moved in direction of force
Power	$\dfrac{\text{work done}}{\text{time taken}}$
Pressure	Force exerted on a unit area

Key ideas

Measuring length, volume and time

- Measure length by looking perpendicularly to the ruler to avoid parallax (see Figure 1.1).

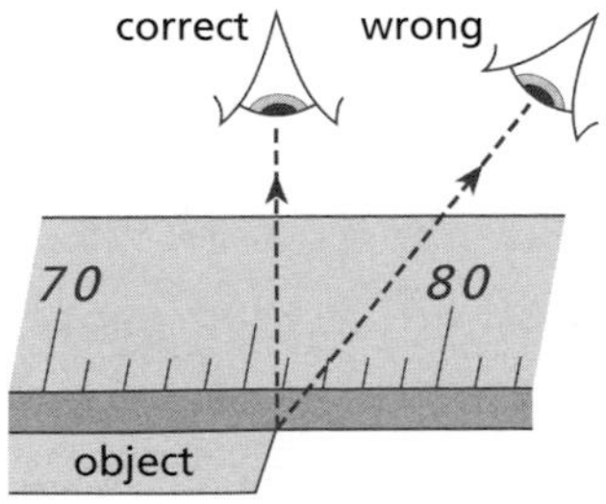

Figure 1.1 The correct way to measure with a ruler

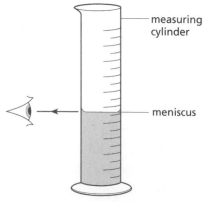

Figure 1.2 The correct way to measure the volume of liquid

- Measure the volume of a liquid by looking level with the bottom of the meniscus (see Figure 1.2). (For mercury, look level with the *top* of the meniscus.)

Common error

✗ Measuring volume of liquid from the top of the meniscus. ■

- Use the start and stop buttons on a stop watch or a clock to measure time **interval**.
- Measure a short repeated time interval by timing a number of cycles and then dividing the total time by the number of cycles.

Common error

✗ Doing the division the wrong way round, that is calculating the repeated time interval as $\dfrac{\text{number of cycles}}{\text{total time}}$ ▨

● **Try this** The answers are given on **p. 103**.

1 A student uses a stop watch to time the swing of a pendulum. He forgets to zero the timer which reads 0.5 s when he starts. He starts the stop watch at the end of the first swing of the pendulum and stops the watch at the end of the tenth swing. The final reading on the timer is 5.9 s. Work out **a)** the number of swings he has timed, **b)** the time taken for these swings, **c)** the time for each swing.

Speed, velocity and acceleration

- Use the formula: $\text{speed} = \dfrac{\text{distance}}{\text{time}}$

Sample question A runner completes an 800 m race in 2 min 30 s after completing the first lap of 400 m in 1 min 10 s. Find her speed for the last 400 m.

[3 marks]

Student's answer Speed $= \dfrac{400}{150} = 2.67$ m/s

[2 marks]

Examiner's comments *The student used the correct formula and correct distance, but used the time for the whole race instead of the time for the last 400 m.*

Correct answer Time $= 2$ min 30 s $- 1$ min 10 s $= 1$ min 20 s $= 80$ s

[1 mark]

Speed $= \dfrac{400}{80} = 5.0$ m/s

[2 marks]

- **Velocity** is speed in a specified direction.
- Use the formula: acceleration = $\dfrac{\text{change of velocity}}{\text{time}}$

Examiner's tips
▶ The formula only applies to *constant* acceleration.
▶ You must be able to recognise linear motion for which acceleration is constant.
▶ You must be able to recognise motion for which acceleration is *not* constant.

● **Distance–time graph**

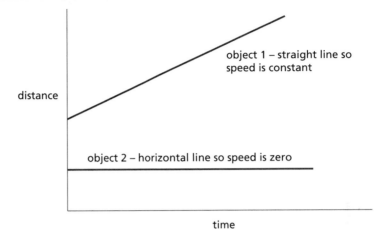

Figure 1.3

● **Speed–time graph**

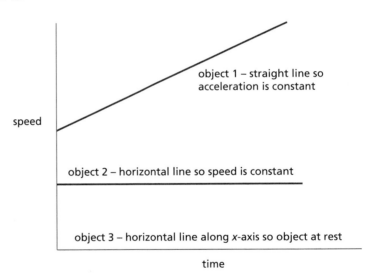

Figure 1.4

- The area under a speed–time graph is the distance covered.
- **Acceleration** occurs when speed changes.
- Beware of the above simplification in the Core syllabus. Supplement students must know the difference between speed and velocity, and that acceleration is rate of change of *velocity*.
- A body in free fall near the Earth has constant acceleration, which is often called *g*.

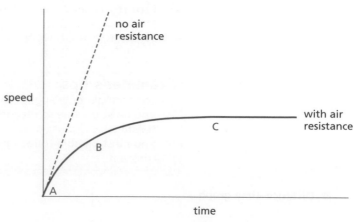

Figure 1.5 A body in free fall in the atmosphere

In the atmosphere there is air resistance. At point A in Figure 1.5, the speed is slow so there is negligible air resistance and the body has free fall acceleration. At point B, the speed is higher and there is some air resistance, so acceleration is less than free fall. At point C, the body has high speed and high air resistance, which is equal to its weight. Therefore, there is no acceleration – this constant speed is called **terminal velocity**.

Common misconception

✗ A body in free fall is weightless.
✓ A body in free fall may feel weightless, but weight is the force of the Earth's gravitational field, which still acts on the body. ■

Sample question A car is moving in traffic and its motion is shown in Figure 1.6.

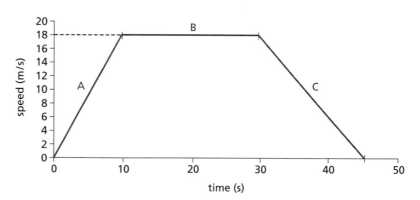

Figure 1.6

a) Choose from the following terms to describe the motion in Parts A, B and C: acceleration, deceleration, steady speed [3 marks]
b) Work out the total distance covered. [5 marks]
c) Work out the acceleration in Part C. [2 marks]

Student's answer a) Part A: acceleration, Part B: deceleration, Part C: steady speed [1 mark]
b) Distance = speed × time = 18 × 45 = 810 m [0 marks]
c) Acceleration = $\dfrac{\text{change of velocity}}{\text{time}} = \dfrac{18}{15} = 1.2$ m/s² [1 mark]

Examiner's comments

a) *The answers to Parts B and C are the wrong way round.*

b) *The formula used is distance = average speed × time, but this is not appropriate as the average speed is unknown. The student should have worked out the area under the graph, which equals the distance covered.*

c) *The calculation is correct but the student should have specified a negative acceleration.*

Correct answer

a) Part A: acceleration, Part B: steady speed, Part C: deceleration   [3 marks]

b) Distance = area under graph   [1 mark]
Part A area = $\frac{1}{2} \times 18 \times 10 = 90\,\text{m}$ [1 mark]
Part B area = $18 \times 20 = 360\,\text{m}$ [1 mark]
Part C area = $\frac{1}{2} \times 18 \times 15 = 135\,\text{m}$ [1 mark]
Distance = total area = 90 + 360 + 135 = 585 m   [1 mark]

c) Acceleration = $\dfrac{\text{change of velocity}}{\text{time}} = \dfrac{-18}{15} = -1.2\,\text{m/s}^2$ [2 marks]

● **Try this**   The answers are given on **p. 103**.

2 A bus accelerates at a constant rate from standstill to 15 m/s in 12 s. It continues at a constant speed of 15 m/s for 8 s.
 a) Show this information on a speed–time graph.
 b) Use the graph to find the total distance covered.
 c) Work out the average speed.

Mass and weight

Examiner's tips
▶ You must be clear about the difference between mass and weight.
▶ Mass measures the amount of matter in an object.
▶ Weight is the force of gravity acting on an object.

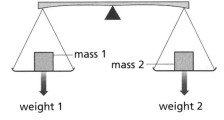

mass 1
mass 2
weight 1
weight 2

Figure 1.7 Balanced weights

• The greater the mass, the more a body resists a change of motion.
• A balance actually compares two weights. As mass determines weight, the balance also compares masses.

In Figure 1.7, mass 1 = mass 2 because weight 1 = weight 2.

Experiments to measure density

Examiner's tips
▶ Measure the mass.
▶ Measure the volume.
▶ Use the formula: density = $\dfrac{\text{mass}}{\text{volume}}$

• For a regularly-shaped solid, measure the dimensions with a ruler and work out the volume. Find the mass on a balance.
• For a liquid, measure the volume in a measuring cylinder. Find the mass of an empty beaker, pour the liquid into the beaker, and find the total mass of the beaker and liquid. Work out the mass of the liquid by subtraction.

Sample question The mass of an empty measuring cylinder is 185 g. When the measuring cylinder contains 400 cm³ of a liquid, the total mass is 465 g. Find the density of the liquid. [4 marks]

Student's answer Density = $\frac{465}{400}$ = 1.16 g/cm³ [2 marks]

Examiner's comments *The student put the appropriate quantities into the correct formula and gave the correct units, but used the total mass instead of working out and using the mass of the liquid itself.*

Correct answer Mass of liquid = 465 − 185 = 280 g [1 mark]

Density = $\frac{280}{400}$ = 0.7 g/cm³ [3 marks]

- For an irregularly-shaped solid, submerge the object in liquid in a large measuring cylinder. The volume of the solid is the increase in the reading (see Figure 1.8).
- Alternatively, use a displacement can. The volume of the solid is the volume of liquid displaced.

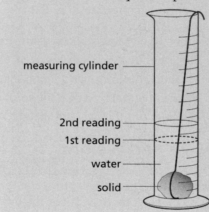

measuring cylinder

2nd reading
1st reading

water

solid

Figure 1.8 Measuring the volume of an irregular solid

● **Try this** The answers are given on **p. 103**.

3 A measuring cylinder containing 20 cm³ of liquid is placed on a top pan balance, which reads 150 g. More liquid is poured into the cylinder up to the 140 cm³ mark and the top pan balance now reads 246 g. A solid is gently lowered into the cylinder; the liquid rises to the 200 cm³ mark and the top pan balance reading to 411 g. Work out **a)** the density of the liquid, **b)** the density of the solid.

Forces and change of size and shape

- Forces can change the size and shape of a body.

Examiner's tips
▶ You must be able to describe an experiment to measure the extension of a spring, piece of rubber or other object with increasing load.
▶ You must be able to plot an extension–load graph with the results of such an experiment.

- Identify the elastic and plastic regions as well as the limit of proportionality of extension–load graphs (see Figure 1.9).

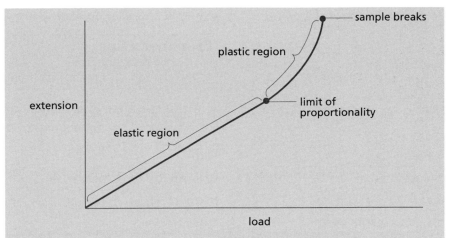

Figure 1.9 Extension–load graph of a typical metal sample loaded to breakage

- State Hooke's Law for the elastic region: extension is proportional to load – the graph is a straight line.
- Recall, understand and use the formula for Hooke's Law: $F = kx$

Forces and change of motion

- A **resultant force** gives an acceleration to an object. If the object is stationary, it will gain speed. If the object is moving, it will gain or lose speed depending on the direction of the force.

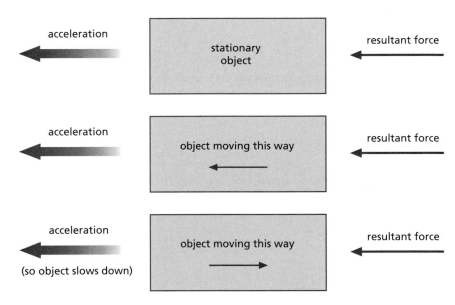

Figure 1.10 A resultant force changes the motion of an object

Common misconception

- ✗ If no forward force acts on a moving body, it will slow down.
- ✓ If a friction force acts on a moving body and there is no forward force, there is a resultant force backwards on the body and it will slow down.
- ✓ If no resultant force acts on a moving body, it will continue moving with the same speed. ∎

- Recall and use the formula: $F = ma$

Examiner's tips
▶ F is the resultant force.
▶ Acceleration a is in the direction of the resultant force.

- When the force is perpendicular to motion, the object follows a circular path. Some examples of this are shown below.

Object	Force	Circular motion
Planet in orbit	Gravitational force towards the Sun	Planet moves around the Sun
Car turning a corner	Friction force	Car drives around the corner
Ball on a length of string	String tension	Ball whirls around in a circle

Examiner's tips
▶ You must be able to find the resultant of two or more forces acting in the same line.
▶ You must state the **direction** of the resultant force.

Sample question An empty lift weighs 2000 N. Four people enter the lift and their total weight is 3000 N. After the button is pressed to move the lift, the tension in the cable pulling up from the top of the lift is 4000 N.

a) Work out the resultant force on the lift. [2 marks]
b) State how the lift moves. [2 marks]
c) Work out the resultant acceleration (take the weight of 1 kg to be 10 N). [4 marks]

Student's answer
a) Resultant force = 3000 + 2000 − 4000 = 1000 N [1 mark]
b) The lift will move down. [1 mark]
c) Mass of lift and people = $\frac{5000}{9.81}$ = 509.7 kg

Acceleration = $\frac{F}{m}$ = $\frac{1000}{509.7}$ = 1.962 m/s² downwards [3 marks]

Examiner's comment
a) *The student correctly worked out the size of the force but did not state the direction downwards.*
b) *The words 'move down' are too vague.*
c) *The student's answer is correct in itself but the correctly remembered exact value for g was used, not the approximate value quoted, thus making life harder! [1 mark is lost for disregard of instructions]*

Correct answer
a) Resultant force = 3000 + 2000 − 4000 = 1000 N downwards. [2 marks]
b) The lift will accelerate downwards. [2 marks]
c) Mass of lift and people = $\frac{5000}{10}$ = 500 kg

Acceleration = $\frac{F}{m}$ = $\frac{1000}{500}$ = 2 m/s² downwards [4 marks]

● **Try this** The answers are given on **p. 103**.

4 A rocket of weight 1000 N is propelled upwards by a thrust of 1800 N. The air resistance is 500 N.
a) Work out the resultant force on the rocket.
b) Describe how this resultant force changes the motion of the rocket.

Turning effect and equilibrium

- The **moment** of a force is its turning effect.

Figure 1.11 Balancing a beam on a pivot

- A beam balances if the anticlockwise moment of the force on the left of the pivot equals the clockwise moment of the force on the right.

- Moment = force × perpendicular distance from pivot

Sample question   A student carries out an experiment to balance a regular 4 m long plank at its mid-point. A mass of 4 kg is placed 80 cm to the left of the pivot and a mass of 3.2 kg is placed 100 cm to the right of the pivot. Explain, *by working out the moments*, whether the plank is balanced. [4 marks]

Figure 1.12

Student's answer   4 × 80 = 3.2 × 100 so the plank balances. [2 marks]

Examiner's comments   *The student's calculation and conclusion are entirely correct but the instruction in italic to work out the moments was ignored.*

Correct answer   Anticlockwise moment = 40 × 0.8 = 32 N m [1 mark]

Clockwise moment = 32 × 1 = 32 N m [1 mark]

Anticlockwise moment = Clockwise moment, so the plank balances. [2 marks]

- An object is in **equilibrium** if there is no resultant turning effect and no resultant force.

● **Try this**   The answer is given on **p. 103**.

5 A see-saw has a total length of 4 m and is pivoted in the middle. A child of weight 400 N sits 1.4 m from the pivot. A child of weight 300 N sits 1.8 m from the pivot on the other side. A parent holds the end of the see-saw on the same side as the lighter child. Work out the magnitude and direction of the force the parent must exert to hold the see-saw level.

Centre of mass and stability

> **Examiner's tip**
> ▶ You must be able to describe and carry out an experiment to find the centre of mass of an irregular, flat object.

• The lower the centre of mass, the more stable an object becomes.

Sample question A girl is seated safely and steadily in a canoe but when she tries to stand up, the canoe capsizes (see Figure 1.13). Explain this in terms of centre of mass and stability. [2 marks]

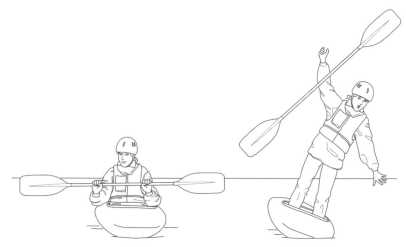

Figure 1.13

Student's answer The centre of gravity goes up so she falls over. [1 mark]

Examiner's comments *The centre of mass does rise but the student did not mention stability. Also, the student used the old term 'centre of gravity' which is not the correct syllabus term, although the student would not lose marks for this.*

Correct answer The centre of mass rises so the canoe becomes unstable and capsizes. [2 marks]

Scalars and vectors

• A **scalar** only has size (magnitude).
• Examples of scalars: mass, speed, energy.
• Scalars are added by normal addition.
• A **vector** has direction and size (magnitude).
• Examples of vectors: force, velocity, acceleration.
• Vectors are added by taking into account their direction.

Common misconception

✗ Speed is a vector.
✓ Speed is a scalar because it has no direction. Velocity has size *and* direction so it is a vector. ■

> **Examiner's tips**
> ▶ You must be able to use a graphical technique to represent two vectors to find their resultant.
> ▶ Learn carefully how to use the parallelogram law *or* the triangle law.

Sample question   An aircraft flies at 900 km/h heading due south. There is a cross-wind of 150 km/h from the west. Graphically, find the aircraft's resultant velocity.   [4 marks]

Student's answer

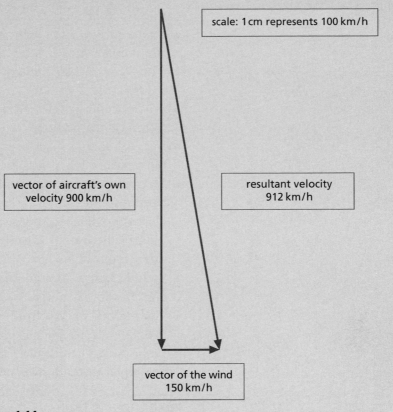

scale: 1 cm represents 100 km/h

vector of aircraft's own velocity 900 km/h

resultant velocity 912 km/h

vector of the wind 150 km/h

Figure 1.14

[3 marks]

Examiner's comments   *On the whole, the question is extremely well-answered and the graphical work accurate; stating the scale shows excellent work. However, the student has omitted the direction part of the resultant velocity, only stating the magnitude.*

Correct answer   The answers shown in Figure 1.14 are correct except that the resultant velocity label should be:

resultant velocity 912 km/h at 9° east of due south   [4 marks]

Energy

You must understand and be able to give examples of the forms of energy listed below.

- Energy due to motion (kinetic energy), e.g. a car moving, a stone falling, a person running.
- Energy due to position (gravitational energy), e.g. water in a mountain lake that can flow downhill to generate electricity, a raised weight that gains gravitational energy due to being in a higher position, energy stored in waves and tides that can also be used to generate electricity.

- Chemical energy released in chemical reactions, e.g. burning fuel to release heat, eating food to provide energy to muscles, providing electrical energy from chemical reactions in a battery, burning fuel in a boiler to provide steam, which can drive a turbine to generate electricity.
- Strain energy due to the stretching or bending of materials, e.g. stretching a rubber band, compressing or extending a spring. The term potential energy includes gravitational and strain energy.
- Nuclear energy released during fission, as in nuclear reactors, or fusion, as in the Sun.
- Internal energy (heat or thermal energy), e.g. the increase in temperature when an object is heated.
- Electrical energy that can be produced in power stations and batteries (but cannot be stored on a large scale), widely-used for electronic devices, motors, lighting and heating in homes and industry because it is easily transmitted and transformed into other types of energy.
- Sound energy – longitudinal pressure waves that travel through a compressible material.
- Geothermal energy from within the Earth, which can be used to generate electricity and provide heat for homes and factories.
- Light energy and other forms of electromagnetic radiation that can travel through a vacuum, e.g. heat and light from the Sun.

Examiner's tips
- ▶ You must be able to give examples of conversion of energy from one form to another.
- ▶ You must be able to give examples of the transfer of energy from one place to another.
- ▶ Energy is *always* conserved. It can be transformed into other forms but cannot be created or destroyed.
- ▶ You must be able to recall and use the formulae: k.e. = $\frac{1}{2}mv^2$ and p.e. = mgh
- ▶ Efficiency is a measure of the proportion of energy input to a device which is output.

$$\text{Efficiency} = \frac{\text{energy output}}{\text{energy input}}$$

It is not essential to know this formula but many students will find it the easiest way to understand the relationship.

- When energy is transformed, some energy is lost to the surroundings as thermal energy, often due to friction (see Figure 1.15).

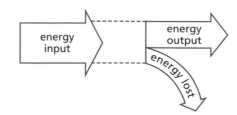

Figure 1.15 Energy transformation involves some loss of heat

Sample question A man winds up the spring of the clockwork radio shown in Figure 1.16 using the muscles in his hand and arm. The internal spring then unwinds to provide energy to power the radio.

Figure 1.16 Winding up a clockwork radio

a) State the type of energy stored in his muscles. [1 mark]

b) State the type of energy stored in the spring. [1 mark]

c) Name the component that converts the energy from the spring into useful energy for the radio. [1 mark]

d) Name the type of energy required by the circuits of the radio. [1 mark]

e) Name the type of useful energy output by the radio. [1 mark]

f) Most of the energy from the spring will eventually be turned into a form of waste energy. Name this type of energy. [1 mark]

Student's answer

a) Chemical energy is stored in his muscles. [1 mark]

b) Potential energy is stored in the spring. [0 marks]

c) A generator converts the energy in the spring. [1 mark]

d) The circuits require electrical energy. [1 mark]

e) The useful output energy is radio waves. [0 marks]

f) The waste energy is friction. [0 marks]

Examiner's comments

a) *Correct answer*

b) *The student should have specified the type of potential energy; a spring stores strain potential energy – often simply called strain energy.*

c) *and* d) *Correct answers*

e) *The radio is a receiver not a transmitter so the output is sound energy.*

f) *Friction is not a type of energy; it is a force which can occur when other types of energy are converted to heat.*

Correct answer

a) Chemical energy is stored in his muscles. [1 mark]

b) Strain (potential) energy is stored in the spring. [1 mark]

c) A generator converts the energy in the spring. [1 mark]

d) The circuits require electrical energy. [1 mark]

e) The useful output energy is sound energy. [1 mark]

f) The waste energy is heat, or thermal energy. [1 mark]

Common misconception

✗ Energy lost as waste heat has been destroyed. ■

● **Try this**

The answers are given on **p. 103**.

6 A bungee jumper of mass 60 kg jumps from a bridge tied to an elastic rope which becomes taut after he falls 10 m. Consider the jumper when he has fallen another 10 m and is travelling at 15 m/s.

State **a)** a form of energy that has been lost, **b)** two forms of energy that have been gained.

c) Work out how much energy is stored in the rope. Take $g = 10$ m/s^2 and ignore air resistance.

Work and power

- **Work** is done when a force moves though a distance.
- The greater the force, the more work is done.
- The greater the distance moved, the more work is done.
- ★ Work done = force × distance moved in direction of force ★
- When work is done energy is transferred.
- You must be able to recall and use the equation: $\Delta W = Fd = \Delta E$, where ΔW is work done and ΔE is energy transferred.
- **Power** is the rate of doing work.
- The greater the work done in a given time, the greater the power.
- The shorter the time in which a given amount of work is done, the greater the power.
- ★ Power $= \dfrac{\text{work done}}{\text{time taken}}$ ★
- You must be able to recall and use the equation: $P = \dfrac{E}{t}$

Examiner's tip
▶ Core students must be able to explain the relationships for work and power. You may find the easiest way to do this is to use the equations marked ★ although the syllabus does not actually require you to know these equations.

Sample question

The two cranes shown in Figure 1.17 are lifting loads at a port. Crane A raises a load of 1000 N through a height of 12 m in 10 s. Crane B raises the same load of 1000 N through the same height of 12 m but takes 12 s.

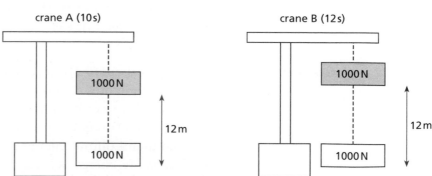

Figure 1.17

a) Compare, with reasons, the work done by the two cranes.

[2 marks]

b) Compare, with reasons, the power of the two cranes.

[2 marks]

c) Calculate the energy transferred and power of each crane.

[4 marks]

Student's answer

a) Both cranes do the same amount of work because the force and distance moved are the same. [2 marks]

b) Crane B has more power because the amount of work done is the same but the time is bigger. [0 marks]

c) Energy transferred by each crane = 1000 × 12 = 12 000

[1 mark]

Power of A = 12 000 × 10 = 120 000 [0 marks]

Power of B = 12 000 × 12 = 144 000 [0 marks]

Examiner's comments

a) *Correct answer*

b) *The student has confused the relationship; the shorter the time taken the greater the power.*

c) *The calculation of energy transferred is correct, except that the unit (J) has been omitted. Both power calculations are incorrect because the wrong equation has been used; the unit of power (W) has also been omitted.*

Correct answer

a) Both cranes do the same amount of work because the force and distance moved are the same. [2 marks]

b) Crane A has more power because the amount of work done is the same but less time is taken. [2 marks]

c) Energy transferred by each crane = 1000 × 12 = 12 000 J

[2 marks]

Power of A = $\frac{12\,000}{10}$ = 1200 W [1 mark]

Power of B = $\frac{12\,000}{12}$ = 1000 W [1 mark]

Pressure

- **Pressure** on a surface is the force exerted on a given area.
- The greater the force on a given area, the greater the pressure.
- The smaller the area on which a given force acts, the greater the pressure.

Examiner's tip

▶ Many Core students may find the easiest way to explain the relationships for pressure is to use the equation: pressure = $\frac{\text{force}}{\text{area}}$ The syllabus does not actually require you to know this equation.

Common error

✗ Carelessly using the word 'pressure' instead of 'force' in the answer to descriptive questions. ■

- You must be able to recall and use the equation: $p = \frac{F}{A}$

- The unit of pressure is the pascal (Pa). A force of 1 N on an area of 1 m^2 exerts a pressure of 1 Pa.

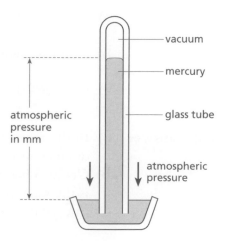

Figure 1.18 Simple mercury barometer

A barometer is used to measure atmospheric pressure in millimetres of mercury, which is given by the height of the mercury column (Figure 1.18).

- Pressure beneath a liquid surface depends on the depth and the density of the liquid.
- The greater the depth in a given liquid, the greater the pressure.
- At a given depth, the greater the density of the liquid, the greater the pressure.
- You must be able to recall and use the equation: $p = \rho g h$, where ρ is the density of the liquid, g the acceleration due to gravity and h the depth below the surface of the liquid.

The manometer shown in Figure 1.19 is used to compare the pressure of a gas with atmospheric pressure. The difference in pressure is measured by the height difference h between the two columns (Figure 1.19).

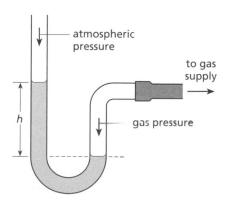

Figure 1.19 Manometer or U-tube filled with water

Sample question Some students are playing a ball game in the sea and the ball is pushed 60 cm below the surface of the water (see Figure 1.20). (Density of sea water = 1.03 × density of fresh water.)

a) Compare, with reasons, the pressure on a point on the ball 60 cm below the surface of the sea with the pressure just below the surface.

[2 marks]

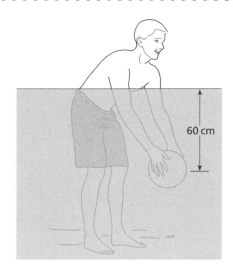

Figure 1.20

b) Compare, with reasons, the pressure on the ball 60 cm below the surface of the sea with the pressure 60 cm below the surface of a fresh water lake. [2 marks]

c) Calculate the pressure on a point on the ball 60 cm below the surface of the sea (density of fresh water = 1000 kg/m³; take *g* to be 10 m/s²). [2 marks]

Student's answer **a)** The pressure increases. [1 mark]

b) The pressure on the ball below the surface of the sea is greater because sea water has a greater density. [1 mark]

c) Pressure = ρgh = 1000 × 10 × 0.6 = 6000 Pa [1 mark]

Examiner's comments **a)** *The statement is correct but no reason is given.*

b) *The statement is correct and the reason is also correct, but not quite complete. The student should have mentioned that the comparison was at the same depth below each surface.*

c) *The pressure has been calculated in the correct way but at a depth of 60 cm below the surface of fresh water instead of sea water.*

Correct answer **a)** The pressure increases because the ball is at a greater depth in the same liquid. [2 marks]

b) The pressure on the ball below the surface of the sea is greater because sea water has a greater density and both balls are at the same depth. [2 marks]

c) Density of sea water = 1.03 × 1000 kg/m³ = 1030 kg/m³
Pressure = ρgh = 1030 × 10 × 0.6 = 6180 Pa [2 marks]

● **Try this**   The answers are given on **p. 103**.

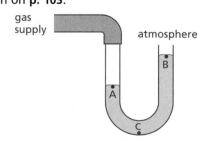

Figure 1.21

7 An engineer uses the mercury manometer shown in Figure 1.21 to test the pressure of a gas supply. Write down at which of the points A, B or C
a) the pressure is greatest, **b)** the pressure is smallest.

TOPIC 2 Thermal Physics

Key objectives

- To know how the kinetic theory explains the nature of solids, liquids and gases
- To know how the flow of thermal energy affects the properties of solids, liquids and gases and their changes of state
- To know how thermal energy is transferred by conduction, convection and radiation
- To be able to explain the behaviour of solids, liquids and gases in a wide range of practical situations from everyday life

Key definitions

Term	Definition
Molecule	A tiny particle consisting of one, two or more atoms
Particle	Any small piece of a substance; could be a molecule or billions of molecules
Temperature	How hot a body is
Thermal energy	Energy that flows into or out of a body by conduction, convection or radiation

Key ideas

Kinetic theory of molecules

- All matter is made up of molecules in motion.
- The higher the temperature, the faster the motion of the molecules.
- Almost always, matter expands with increase of temperature.

Solids

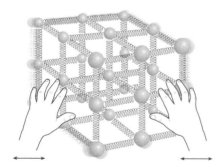

Figure 2.1 A model of molecular behaviour in a solid

- Molecules are close together.
- Molecules vibrate about fixed points.
- The rigid structure of solids results from these fixed positions.
- As the temperature increases, the molecules vibrate further and faster. This pushes the fixed points further apart and the solid expands.
- The positions of molecules in a solid are fixed because the attractive and repulsive forces between neighbouring molecules are balanced.
- There is only a very slight expansion of a solid with increase of temperature, e.g. the length of an iron rod increases by about 0.1% when it is heated from 20 °C to 100 °C.

Liquids

- Molecules are slightly further apart than in solids.
- Molecules are still close enough to keep a definite volume.
- Molecules move randomly in all directions, not being fixed to each other.

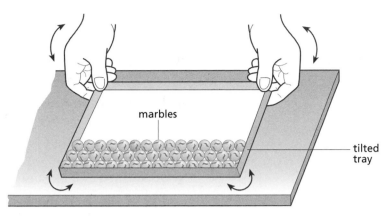

Figure 2.2 A model of molecular behaviour in a liquid

- As the temperature increases, the molecules move faster and further apart so the liquid expands. One *exception* to this is: liquid water heated from 0 °C to 4 °C changes its structure so it contracts instead of expands.
- The forces between molecules are too weak to keep them in a definite pattern but are sufficient to hold them to the bulk of the liquid.
- There is a small expansion of a liquid with increase of temperature, e.g. the volume of many liquids increases by about 4% when heated from 20 °C to 100 °C.

Everyday uses and consequences of thermal expansion

Examiner's tip
▶ You must be able to state and explain some of the everyday uses and consequences of thermal expansion. As a minimum, learn one constructive use and one disadvantage each for solids and liquids.

- Uses of expansion of a solid: shrink-fitting, curling of a bimetallic strip in a fire alarm.

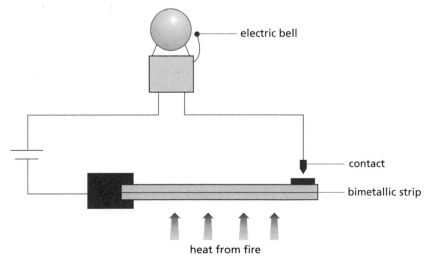

Figure 2.3 A fire alarm

In the fire alarm circuit of Figure 2.3, the heat from the fire causes the lower metal in the bimetallic strip to expand more than the upper metal. This causes the strip to curl up, which completes the circuit and the alarm bell rings.

- Disadvantage of expansion of a solid: gaps need to be left between lengths of railway line to allow for expansion in hot weather.
- Use of expansion of a liquid: a mercury or alcohol thermometer; see the later section on **Measurement of temperature**.
- Disadvantage of expansion of a liquid: the water in a car's cooling system expands when the engine gets hot. A separate water tank is needed for the hot water to expand into.

Sample question The lid is stuck on a jam jar. How could you use hot water to release it? Explain in terms of the molecules how this works.

[4 marks]

Student's answer Put the jam jar in hot water and the lid will come off [1 mark] because the molecules expand. [0 marks]

Examiner's comments *The student did not specify where exactly the hot water should be used and gave a vague, incorrect explanation of the role of the molecules.*

Correct answer Put the lid in hot water so it expands and can be released. [2 marks] The molecules in the lid will move faster; their mean positions move further apart so the lid expands. [2 marks]

● **Try this** The answer is given on p. 103.

1 Two straight strips of metal alloys, invar and bronze, are bonded together at room temperature. Bronze expands appreciably when heated but invar expands very little. Describe the shape of the strips when heated in an oven.

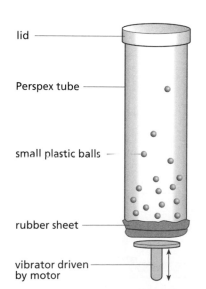

lid

Perspex tube

small plastic balls

rubber sheet

vibrator driven by motor

Figure 2.4 A model of molecular behaviour in a gas

Gases
- Molecules are much further apart than in solids or liquids.
- Molecules move much faster than in solids or liquids.
- There is no definite volume. Molecules move throughout the available space.
- Molecules constantly collide with each other and the container walls.
- Gases have low densities.
- The higher the temperature, the faster the speed of the molecules. In fact, temperature is a measure of the average speed of the molecules.
- The higher the temperature, the larger the volume of a gas at constant pressure.
- Except when they are actually colliding, the forces between the molecules are negligible.
- There is a considerable expansion of a gas with increase of temperature at constant pressure, e.g. the volume of a gas increases by about 27% when it is heated from 20 °C to 100 °C.

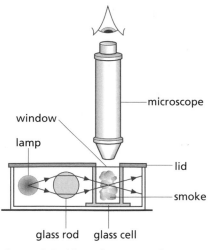

Figure 2.5 Observing Brownian motion

Brownian motion

- In the apparatus in Figure 2.5, smoke particles reflect the light, which is seen in the microscope as bright pin pricks.
- They move around randomly. They also move in and out of focus as they move vertically.
- This movement is caused by the irregular bombardment of the smoke particles by fast-moving, **invisible** air molecules.
- This is clear evidence for the kinetic theory.
- This was first observed by Robert Brown who observed in a microscope pollen particles suspended in water moving due to bombardment by fast-moving, invisible water molecules.
- Smoke/Pollen particles are visible and relatively massive.
- Air/Water molecules are invisible and fast-moving.

Common error

✗ Incorrect use of the words 'particle' and 'molecule' shows a lack of understanding of their different properties and roles. ∎

Sample question A student looks in a microscope at a cell containing illuminated smoke particles. Explain **a)** what is seen, **b)** the movement observed, **c)** what causes this movement. [4 marks]

Student's answer
a) smoke particles [0 marks]
b) moving around [0 marks]
c) The smoke molecules are bombarded by air. [1 mark]

Examiner's comments
a) *It is reflected light, not smoke particles that are seen.*
b) *'Moving around' is too vague.*
c) *The student has made incorrect/incomplete use of the terms 'molecule' and 'particle'.*

Correct answer
a) bright specks of light [1 mark]
b) moving around haphazardly in *all* directions [1 mark]
c) The bright specks are light reflected off the smoke particles, which are bombarded by air molecules. [2 marks]

Measurement of temperature

- A physical property is needed that varies in a regular way over a wide range of temperatures. A practical thermometer is a simple piece of apparatus that measures how this property changes.
- Examples of properties that can be used: expansion of a solid (coiled bimetallic strip) or liquid (mercury-in-glass), electrical voltage between two junctions of different metals (thermocouple), electrical resistance.

Common errors

✗ Confusing heat and temperature.
✓ Temperature measures how hot a body is. Heat is energy that flows from a hot body to a cold body. ∎

• The thermometer must have sufficient **sensitivity**. This means the property must change enough to be measurable, e.g. enough coils in a coiled bimetallic strip thermometer to give sufficient movement, a narrow tube in a liquid-in-glass thermometer.

• The thermometer must have sufficient **range**. This means it can be used over a wide range of temperatures, e.g. apart from other problems, a water-in-glass thermometer would be of limited use because it could only be used between 0 °C and 100 °C.

• The thermometer's reading must show **linearity**. This means it must change by the same amount for every degree of temperature change, e.g. a property that changes little in one half of the temperature range and much more in the other half would not be suitable.

• The **fixed points** of a thermometer are essential to give it its scale. The thermometer must read exactly 0 °C at the freezing point of pure water, and exactly 100 °C at the boiling point of pure water, at normal atmospheric pressure. In between these fixed points the scale is divided into equal divisions.

Liquid-in-glass thermometers

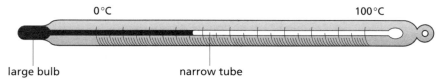

Figure 2.6 A mercury thermometer with a Celsius scale

The liquid only expands by a few per cent, so the expansion of the large amount of liquid into the narrow tube gives a good amount of movement, which can be easily read. Mercury and alcohol are suitable liquids because:

1) they expand enough to make a sensitive thermometer
2) they can be used over wide ranges of temperatures – alcohol from −115 °C to 78 °C (and higher if under pressure) and mercury from −39 °C to 357 °C.
3) they both expand linearly with increase of temperature.

● **Try this**   The answers are given on **p. 103**.

2 A mercury-in-glass thermometer is placed in pure melting ice and the mercury bead is 12 mm long.
  a) What Celsius temperature should the thermometer read?
  The thermometer is now placed in steam above boiling water and the bead expands to 82 mm long.
  b) What Celsius temperature should the thermometer now read?
  c) Work out the length of the bead at 50 °C.
  d) Work out the temperature reading when the bead length is 61 mm.

Thermocouple

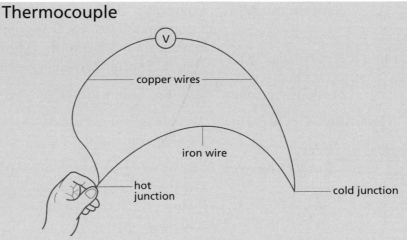

Figure 2.7 A simple thermocouple thermometer

The voltage produced between the two junctions of wires of different metals is proportional to the temperature difference between the junctions. Only the hot junction is attached to the object whose temperature is being measured. This junction is very small and light, so reacts quickly to rapidly changing temperatures. It is also very robust and resistant to damage from vibrations of machinery. Thermocouples are widely used in industry because they can be used over a wide temperature range from below −250 °C to 1500 °C. In addition, the voltage output is very convenient for data logging; an industrial test rig may have hundreds of thermocouple thermometers all connected to a computer for analysis.

Sample question A new petrol engine is being tested and the engineers need to measure the temperature of the exhaust pipe close to the engine. State with reasons the type of thermometer that would be used.

[4 marks]

Student's answer A thermocouple because it is small and goes up to a high temperature. [2 marks]

Examiner's comments *The student's answer is on the right lines but rather vague and incomplete. (It would earn the first and last marks below.)*

Correct answer A thermocouple would be used [1 mark] because the hot junction is small [1 mark] and will not be damaged by the vibrations of the engine. [1 mark] A thermocouple can measure high temperatures. [1 mark]

Changes of state

- Molecules at the surface of a liquid, which are moving fastest, have enough energy to escape the attractive force of the rest of the liquid and **evaporate** to become molecules of a gas.
- Evaporation takes place at all temperatures.
- Energy is needed to break the bonds between molecules so evaporation causes the remaining liquid to cool down.

Sample question A student is playing football on a cool, windy day, wearing a T-shirt and shorts. He feels comfortably warm because he is moving around vigorously. His kit then gets wet in a rain shower. Explain why he now feels cold. [2 marks]

Student's answer The wet T-shirt makes him feel cold. [0 marks]

Examiner's comments *The student's answer is far too vague and does not mention the cooling caused by evaporation.*

Correct answer The water in his wet kit is evaporated by the wind. [1 mark] The thermal energy needed for this evaporation is taken from the water in his T-shirt and shorts as well as from his body so he feels cold. [1 mark]

● **Try this** The answers are given on **p. 103**.

3 An ice cube with a temperature of 0 °C is placed in a glass of water with a temperature of 20 °C. After a few minutes some of the ice has melted. State whether the following increase, decrease or stay the same: **a)** temperature of the ice, **b)** temperature of the water, **c)** mass of the water, **d)** total mass of the ice and water.

- Rate of evaporation increases with:
  1) higher temperature, as more molecules at the surface are moving faster
  2) increased surface area, as more molecules are at the surface
  3) a wind or draught, as the gas molecules are blown away so cannot re-enter the liquid.

- **Boiling** occurs at a definite temperature called the 'boiling point'. Bubbles of vapour form within the liquid and rise freely to the surface. Energy must be supplied continuously to maintain boiling.

Examiner's tip
▶ You must be able to distinguish between **boiling** and **evaporation**. Carefully learn the features of each.

- **Condensation** occurs when gas or vapour molecules return to the liquid state. Energy is given out as the bonds between molecules in the liquid re-form.
- **Melting**, or fusion, takes place at a definite temperature called the 'melting point'. Energy must be provided to break the bonds between molecules in order for them to leave the well-ordered structure of the solid.
- **Solidification**, or freezing, occurs when molecules of a liquid return to the solid state. This takes place at a definite temperature called the 'freezing point', which has the same value as the melting point. Energy is given out as the bonds between molecules of the solid re-form.

Common misconception

✗ Temperature increases during melting and boiling because thermal energy is being supplied.

✓ Temperature stays constant during melting and boiling. The thermal energy supplied goes to break the bonds between molecules. ■

- Energy must be provided to change state from solid to liquid, or liquid to gas. This energy is called 'latent heat' and is used to break bonds between molecules, not to increase temperature.
- **Specific latent heat of fusion** is defined as the thermal energy needed per kilogram to change a material from solid to liquid, at constant temperature.

● An experiment to measure the specific latent heat of fusion of ice

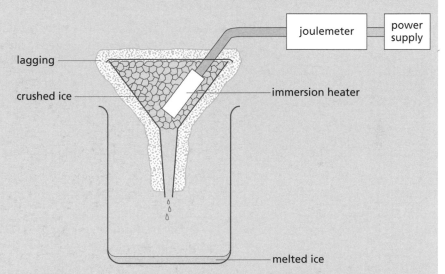

Figure 2.8 The lagging reduces melting caused by heat from the surroundings

Measure the mass of melted ice from the funnel and record the joulemeter reading of the energy supplied by the immersion heater. Use these results to calculate the specific latent heat of fusion.

Example of working out:

Mass of beaker = 80 g
Mass of beaker and melted ice = 195 g

Mass of ice melted by immersion heater = 195 − 80 = 115 g
= 0.115 kg

Joulemeter reading = 45 000 J

$$\text{Specific latent heat of fusion} = \frac{\text{energy supplied}}{\text{mass of ice melted}}$$

$$= \frac{45\,000}{0.115} = 391\,000\,\text{J/kg}$$

- **Specific latent heat of vaporisation** is defined as the thermal energy needed per kilogram to change a material from liquid to gas, at constant temperature.

● **An experiment to measure the specific latent heat of vaporisation of water**

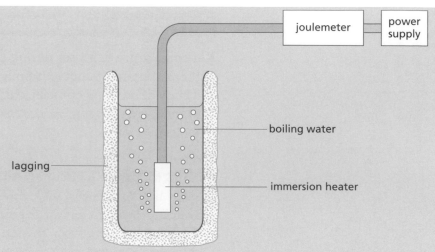

Figure 2.9 The lagging and tall, narrow shape reduce heat losses to the surroundings

Measure the mass of water and its container before and after the experiment. Bring the water to the boil before starting the joulemeter. Record the joulemeter reading of the energy supplied during boiling. Use these results to calculate the specific latent heat of vaporisation.

Example of working out:

Initial mass of water and container = 650 g
Final mass of water and container = 580 g

Mass of water vaporised by immersion heater
   = initial mass of water and container − final mass of water and container
     = 650 g − 580 g = 70 g = 0.07 kg
Joulemeter reading = 245 000 J

$$\text{Specific latent heat of vaporisation} = \frac{\text{energy supplied}}{\text{mass of water vaporised}}$$

$$= \frac{245\,000}{0.07} = 3\,500\,000\,\text{J/kg}$$

Sample question

Ice is melted in a lagged funnel by an immersion heater supplied with electricity through a joulemeter. The melted water is collected in a beaker. Use the data supplied to work out the specific latent heat of fusion of ice.

Joulemeter readings: before experiment 148 000 J
                           after experiment 172 600 J
Mass of empty beaker: 90 g
Mass of beaker with melted water: 150 g   [4 marks]

Student's answer

Heat emitted by immersion heater = 172 600 − 148 000 = 24 600 J   [1 mark]

Specific latent heat of fusion = $\dfrac{24\,600}{0.150}$ = 164 000 J/kg   [1 mark]

Examiner's comments

The student has made a good attempt at the calculation but forgot to subtract the mass of the beaker itself.

Correct answer

Heat emitted by immersion heater = 172 600 − 148 000 = 24 600 J   [1 mark]

Mass of ice melted = 150 − 90 = 60 g = 0.060 kg   [1 mark]

Specific latent heat of fusion = $\dfrac{24\,600}{0.060}$ = 410 000 J/kg   [2 marks]

Gas pressure

- Gas pressure is caused by the total force of collisions between fast-moving molecules and the walls of the container they are in.
- The higher the temperature, the faster the molecules move. If the volume is kept constant, the pressure increases because:
  1) there are more frequent collisions with the container walls
  2) the collisions are harder so exert more force.

Common error

✗ Mentioning collisions between molecules when explaining gas pressure. It is true that the molecules collide with each other but this does not explain gas pressure. ■

Gas pressure and volume at constant temperature

- At a **constant temperature**, gas molecules move at a constant average speed, so the force from each collision is the same (on average).
- If the gas is compressed into a smaller volume, there are more frequent collisions on each unit of area, so the total force per unit area increases and the pressure increases.
- Similarly, if the gas expands to a greater volume at a constant temperature, the pressure decreases.
- At a constant temperature: pressure × volume = constant

> **Examiner's tips**
> ▶ You must be able to solve problems using the equation above; below is a worked example.
> ▶ If you set out your working logically, as shown below, you are much more likely to get the answer right.

Worked example

Question

A piston slowly compresses a gas from $540\,\text{cm}^3$ to $30\,\text{cm}^3$, so that the temperature remains constant. The initial pressure was $100\,\text{kPa}$; find the final pressure.

Solution

$p_1 = 100\,\text{kPa}$, $p_2 = \text{unknown}$

$V_1 = 540\,\text{cm}^3$, $V_2 = 30\,\text{cm}^3$

$p_1 V_1 = p_2 V_2$

$100 \times 540 = p_2 \times 30$

$p_2 = \dfrac{100 \times 540}{30}$

$\quad = \dfrac{54\,000}{30} = 1800\,\text{kPa}$

Sample question A gas cylinder is heated in a fire. State what happens to the pressure of the gas and explain your answer in terms of the molecules. [4 marks]

Student's answer The pressure increases because the molecules move around more, hitting each other and the walls. [1 mark]

Examiner's comments *The student's answer is vague, mentioning the molecules colliding with each other which is irrelevant.*

Correct answer The pressure increases because the molecules move faster, [2 marks] hitting the walls more frequently and harder, thus increasing the total force on the walls. [2 marks]

● **Try this** The answers are given on **p. 103**.

4 An experiment is carried out on some gas contained in a cylinder by a piston which can move.

In Stage 1, the gas is heated with the piston fixed in position. State whether the following increase, decrease or stay the same during Stage 1: **a)** speed of molecules, **b)** number of collisions per second between molecules and the walls, **c)** gas pressure.

In Stage 2, the gas stays at a fixed temperature while the piston moves to increase the volume of the gas. State whether the following increase, decrease or stay the same during Stage 2: **a)** speed of molecules, **b)** number of collisions per second between molecules and the walls, **c)** gas pressure.

Thermal capacity

- When thermal energy flows into a body its molecules move faster, increasing its internal energy and its temperature.
- The **thermal** (or **heat**) **capacity** of a body is the amount of thermal energy needed to increase its temperature by 1 °C. The thermal capacity is determined by the mass of the body and its material.

Common misconception

✗ A material has a thermal (or heat) capacity.
✓ An object or a certain mass of a material has a thermal (or heat) capacity. ■

- **Specific heat capacity** is defined as the thermal energy needed *per kilogram* to increase the temperature of a material by 1 °C.

Common misconception

✗ An object has a specific heat capacity.
✓ Specific heat capacity is a property of a material. ■

An experiment to measure the specific heat capacity of a metal

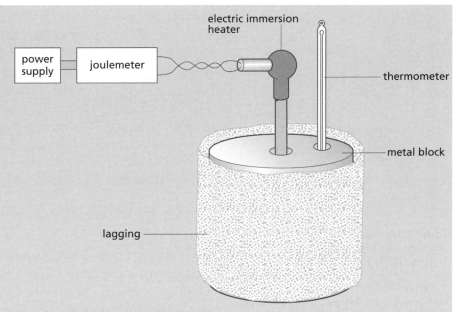

Figure 2.10 The lagging reduces heat losses to the surroundings

Measure the mass of the metal block and the temperature before a and after heating, and record the joulemeter reading of the energy supplied. Use these results to calculate the specific heat capacity of the block.

Example of working out:

Mass of metal block = 1.6 kg
Temperature before heating = 21 °C
Temperature after heating = 66 °C
Increase of temperature = 45 °C
Joulemeter reading = 46 800 J

$$\text{Specific heat capacity} = \frac{\text{energy supplied by immersion heater}}{\text{mass} \times \text{temperature increase}}$$

$$= \frac{46\ 800}{1.6 \times 45} = 650\,\text{J/kg}\,°\text{C}$$

Try this
The answers are given on **p. 103**.

5 An experiment is carried out to find the specific heat capacity of a metal. A 2 kg block of the metal is heated by a 200 W heater for 5 minutes, and the temperature of the block rises from 20 °C to 51 °C. Work out **a)** the energy supplied to the block by the heater, **b)** the specific heat capacity of the metal.
  When used in an engine, a component made from this metal receives 35 kJ of thermal energy and its temperature rises from 30 °C to 290 °C. **c)** Work out the mass of the component.

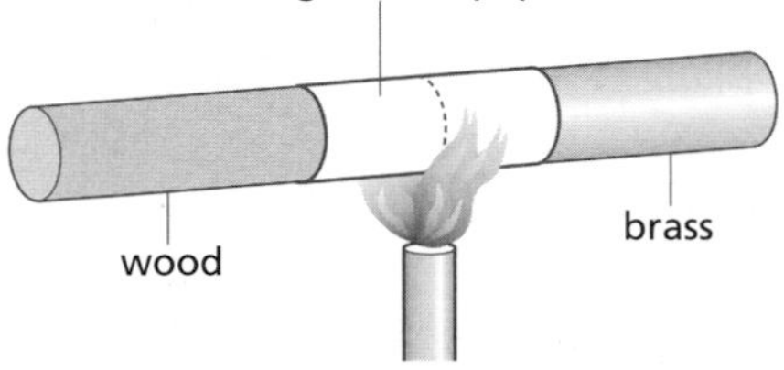

Transfer of thermal energy
Transfer by conduction

- Heat is always transferred from a place of high temperature to a place of low temperature. In **conduction**, heat is transferred through a material without movement of the material.
- Metals are generally good conductors but most other solids are poor conductors.

Figure 2.11 The paper over the brass does not burn because brass is a good conductor

white gummed paper

wood

brass

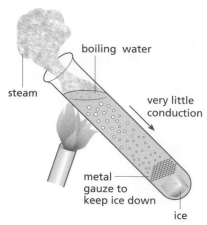

Figure 2.12 Water is a poor conductor of heat

- Liquids are generally much worse heat conductors than metals.
- Gases are all very poor conductors of heat, e.g. if you put your hands in ice water they will feel very cold almost at once. If your hands are in air of the same temperature, they will cool down but at a much slower rate because air is a bad conductor.
- The molecules in a hot part of a solid vibrate faster and further than those in a cold part. Energy is passed through the solid by these faster-moving molecules causing their neighbours to vibrate more. This continues until the molecules in the coldest part also vibrate more and are thus at a higher temperature.

Transfer by convection

- Heat is always transferred from a place of high temperature to a place of low temperature. In **convection**, heat is transferred by a fluid due to movement of the fluid itself.
- The fluid expands on heating so its density falls. The warmer and lighter fluid rises to the cooler region, transferring heat in the process.

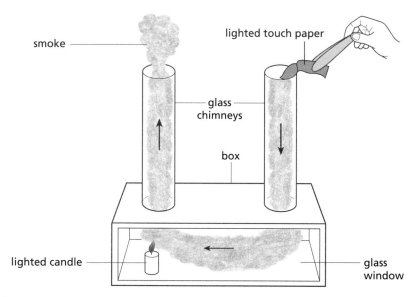

Figure 2.13 Demonstrating convection in air

The convection current from the heat of the candle rises up the left-hand chimney and draws smoke from the lighted paper down the right-hand chimney and into the box.

- Convection currents in water can be seen by dropping a potassium permanganate crystal into a beaker of water. The coloured traces indicate the flow of the convection currents.

Transfer by radiation

- Heat is always transferred from a place of high temperature to a place of low temperature. In **radiation**, heat is transferred by infra-red radiation, which is part of the electromagnetic spectrum.
- There need not be any matter between the hot and cold bodies.
- Most solids and liquids absorb infra-red radiation, including water which is transparent to light.

● **Good and bad emitters of infra-red radiation**

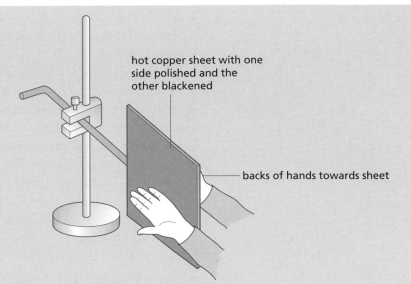

Figure 2.14 Comparing emitters of radiation

The copper sheet has previously been heated strongly with a Bunsen burner. The hand next to the black surface feels much hotter than the hand next to the polished surface. This is because black surfaces emit heat radiation more than polished surfaces.

● **Good and bad absorbers of infra-red radiation**

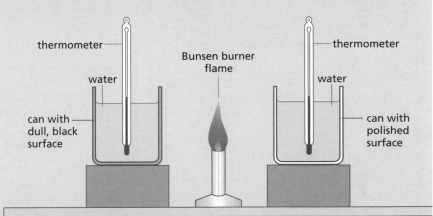

Figure 2.15 Comparing absorbers of radiation

The flame of the Bunsen burner radiates heat. Black surfaces absorb heat radiation more than polished surfaces. The temperature of the water in the can with the blackened surface rises faster than the temperature of the water in the polished can.

● Surfaces that are good absorbers of heat radiation are also good emitters.

Everyday uses and consequences of the transfer of thermal energy

Examiner's tip

▶ There are countless everyday uses and consequences of the transfer of thermal energy. As a minimum, you should learn and be able to explain the ones following. If possible, read up and understand many more, especially coastal breezes, the vacuum flask and the greenhouse.

- Saucepans and other solids through which heat must travel are made of metals such as aluminium or copper, which are **good conductors**.
- Blocks of expanded polystyrene are used for house insulation because they contain trapped air, which is a **bad conductor**.
- A domestic radiator heats the air next to it which then rises and transfers heat to the rest of the room. Despite its name, a radiator works mainly by **convection**.
- Double-glazing reduces heat losses by trapping a narrow layer of air between the window panes and **reducing convection**.
- The Sun heats the Earth by **radiation** through space.
- The cooling fins at the back of a refrigerator are painted black to increase heat loss by radiation because black surfaces are **good emitters**.
- Many buildings in hot countries are painted white because white surfaces are **bad absorbers**.

Common misconception

✗ Black surfaces affect conduction and convection.
✓ The type of surface only influences radiation. ■

Sample question

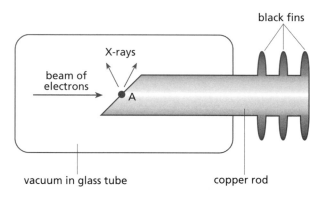

Figure 2.16

Figure 2.16 shows an X-ray tube. Only a small proportion of energy from the electrons that strike Point A goes into the X-rays that are emitted from that point. Most of the energy becomes heat produced at Point A, which is removed by the copper rod. Explain how conduction, convection and radiation play a role in the removal of this heat.　　　　　[4 marks]

[6 marks]

Student's answer　The heat goes down the rod and is conducted out by the fins [1 mark] through convection and radiation. [1 mark]

Examiner's comments　*The student's answer shows incorrect understanding of conduction.*

Correct answer　Heat is **conducted** along the rod to the fins. [2 marks] Heat is emitted from the fins to the air by **convection** and **radiation**. [2 marks] The black colour of the fins increases the rate of radiation. [2 marks]

● **Try this**   The answers are given on **p. 103**.

6 A boy goes for a walk in winter in a cold country. He opens a metal gate which makes his hands cold.

   a) Which type of heat transfer cools his hands?

   He then washes his hands in a stream.

   b) Which type of heat transfer cools his hands in this case?

   He comes across some workmen who have lit a fire to keep themselves warm and he holds out his hands towards the fire.

   c) Which type of heat transfer warms his hands?

TOPIC 3 Waves

Key objectives

- To know the basic nature of wave motion
- To be able to distinguish between transverse and longitudinal waves
- To be able to describe how water waves can be used to illustrate reflection, refraction and diffraction
- To be able to draw diagrams of light rays undergoing reflection, refraction and passing through thin converging lenses
- To know the main features of the electromagnetic spectrum
- To know that sound waves are longitudinal waves produced by a vibrating source, which require a material for their transmission
- To know and be able to use the wave equation: $v = f\lambda$

Key definitions

Term	Definition
Transverse wave	Travelling wave in which oscillation is perpendicular to direction of travel
Longitudinal wave	Travelling wave in which oscillation is parallel to direction of travel
Speed of a wave	The distance moved by a point on a wave in one second
Frequency	The number of complete cycles which occur in one second
Wavelength	The distance between corresponding points in successive cycles of a wave
Amplitude	The maximum displacement of a wave from the undisturbed position
Refractive index	$\dfrac{\text{speed of wave in medium 1}}{\text{speed of wave in medium 2}}$
Principal focus	The point through which all rays parallel to the axis of a lens are refracted
Focal length	The distance between the centre of a lens and the principal focus
Real image	An image formed on a screen by the intersection of rays
Virtual image	An image seen by observing rays diverging from it

Key ideas

The nature of waves

- Waves transfer energy from one point to another without transferring matter.
- Some waves, e.g. water waves and sound waves, are transmitted by particles of a material vibrating about fixed points. They cannot travel through a vacuum.
- Electromagnetic waves, e.g. light waves and X-rays, are a combination of travelling electric and magnetic fields. They *can* travel through a vacuum.

- In **transverse waves**, the oscillation of the material or field is at right angles to the direction of travel of the wave.
- Figure 3.1 represents a transverse wave travelling in a horizontal spring. Each coil oscillates vertically about a fixed point, but not in time with each other.

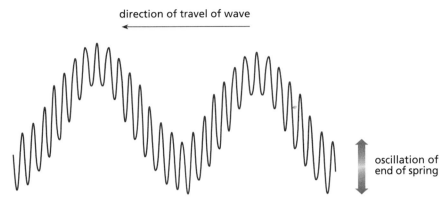

direction of travel of wave

oscillation of end of spring

Figure 3.1 Transverse waves in a spring

Common misconception

✗ Transverse waves oscillate vertically when the wave travels vertically.
✔ Transverse waves oscillate horizontally when the wave travels vertically, because the oscillation is always at right angles to the direction of travel. ■

- In **longitudinal waves**, the oscillation of the material is parallel to the direction of travel of the wave.
- Figure 3.2 represents a longitudinal wave travelling in a horizontal spring. Each coil of the spring oscillates horizontally about a fixed point, but not in time with each other. C marks the points where the coils are most tightly packed (compressions) and R marks the points where the coils are furthest apart (rarefactions).

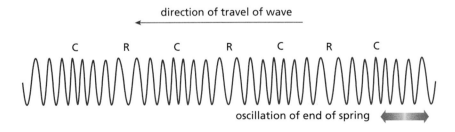

direction of travel of wave

C   R   C   R   C   R   C

oscillation of end of spring

Figure 3.2 Longitudinal waves in a spring

Common misconception

✗ Longitudinal waves oscillate horizontally when the wave travels vertically.
✔ Longitudinal waves oscillate vertically when the wave travels vertically, because the oscillation is always parallel to the direction of travel. ■

- **Wave theory** states that each point on an original wavefront acts as the source of circular, secondary wavelets.

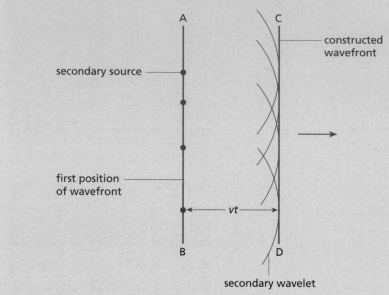

Figure 3.3 Construction for a straight wavefront

The new wavefront is the line touching all the secondary wavelets.

- The speed v of a wave is the distance moved by a point on the wave in one second.
- The **frequency** f of a wave is the number of complete cycles that occur in one second and is measured in hertz (Hz).
- The **wavelength** λ of a wave is the distance between two corresponding points, e.g. crests, in successive cycles.
- The **amplitude** of a wave is the maximum displacement of the wave from the undisturbed position.

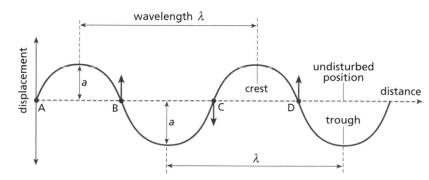

Figure 3.4 Displacement–distance graph for a wave at a particular instant

Sample question Sketch one and a half cycles of a water wave and mark on your sketch the amplitude and wavelength.   [4 marks]

Student's answer The student's answer is shown by dashed lines in Figure 3.5.
          [1 mark]

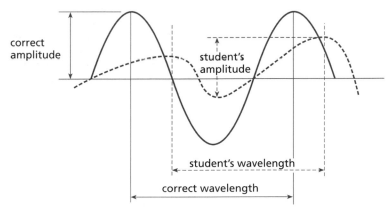

Figure 3.5

Examiner's comments *With this type of question, it is essential to work carefully and accurately or few marks will be gained. Even for a sketch the student's diagram is too casual. For minimum acceptable accuracy, the student must indicate that each half wavelength is the same length, and the distance from the axis to each crest and trough is the same. The student's amplitude is incorrectly measured from crest top to trough bottom. Given the irregular wave, the student has correctly labelled one wavelength.*

Correct answer The correct answer is shown by the solid line in Figure 3.5. [4 marks]

Common misconception

✗ Amplitude is the height difference between the top of a crest and the bottom of a trough. ■

Sample question A sensor detects that 1560 cycles of a wave pass in 30 seconds. Work out the frequency of the wave. [3 marks]

Student's answer Frequency = 52 cycles in one second [2 marks]

Examiner's comments *The student's answer is correct but the unit of frequency is hertz (Hz).*
Note that although it is a fairly easy calculation, there is no working. If the student had made a slight slip, no credit would have been given for using the correct method.

Correct answer Frequency $= \dfrac{\text{number of cycles}}{\text{time}} = \dfrac{1560}{30} = 52\,\text{Hz}$ [3 marks]

● **Try this** The answers are given on **p. 103**.

1 A woman swimming in the sea estimates that when she is in the trough between two waves, the crests are 1.5 m above her. **a)** Work out the amplitude of the waves.
 An observer counts that the swimmer moves up and down 12 times in one minute. **b)** Work out the frequency of the waves.

Examiner's tip
▶ You need to be able to recall and use the wave equation: $v = f\lambda$

Sample question Find the frequency of a radio wave with a wavelength of 1500 m. (Speed of electromagnetic waves $= 3 \times 10^8$ m/s.) [4 marks]

Student's answer	$v = f\lambda$	[1 mark]
	so $f = \dfrac{v}{\lambda} = \dfrac{3 \times 10^8}{1500} = 200\,000$ kHz	[2 marks]
Examiner's comments	*The student has done everything correctly except assume that, as radio frequencies are often expressed in kHz, this was the correct unit.*	
Correct answer	$v = f\lambda$	[1 mark]
	so $f = \dfrac{v}{\lambda} = \dfrac{3 \times 10^8}{1500} = 200\,000$ Hz $= 200$ kHz	[3 marks]

Summary of some types of waves

Type of wave	Longitudinal/Transverse	Travel through a material
Wave on a rope	Transverse	Material needed
Wave on a spring	Either	Material needed
Water	Transverse	Material needed
Earthquake wave	Both	Material needed
Sound	Longitudinal	Material needed
Electromagnetic, e.g. light, X-rays, radio waves	Transverse	No material needed, but some electromagnetic waves can travel through certain materials

Sample question	An earthquake wave is travelling vertically down into the Earth; the oscillations are also vertical. State with a reason whether the wave is longitudinal or transverse. [2 marks]
Student's answer	The wave is transverse because it is vibrating up and down. [0 marks]
Examiner's comments	*It is the direction of oscillation relative to the direction of travel that matters – the student does not mention this.*
Correct answer	The wave is longitudinal because the oscillations are parallel to the direction of travel. [2 marks]

● **Try this** The answers are given on **p. 103**.

2 A child throws a ball into a pond, hears the sound of the splash and observes water waves travelling towards him.
   a) As the sound waves travel towards him, in which direction are the air particles oscillating?
   b) As the water waves travel towards him, in which direction are the water particles oscillating?

Examiner's tip
▶ You must be able to describe how water waves can be used to show reflection, refraction and diffraction.

Water waves

- We can observe waves travelling on the water surface of a ripple tank to illustrate how waves behave.
- Key features of a ripple tank:
  1) A beam just touching the surface vibrates vertically to produce waves.
  2) A light source shines through the water and shows the wave pattern on a screen above or below the ripple tank.

Reflection

In Figure 3.6, the waves produced by the vibrating beam are reflected from the flat metal barrier. The reflected waves are at the same angle to the reflecting surface as the incident waves. Speed, wavelength and frequency are unchanged by reflection.

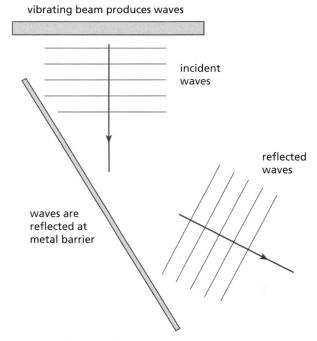

Figure 3.6 Reflection of waves in a ripple tank

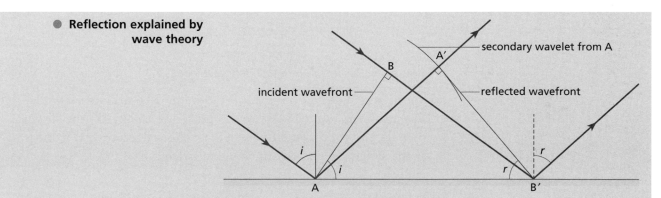

● **Reflection explained by wave theory**

Figure 3.7 Reflection of a straight wavefront

Speed and wavelength stay constant after reflection. The distance travelled by the secondary wavelet AA' equals BB', the distance travelled by the incident wavelet. So the reflected wavefront is at the same angle to the reflecting surface as the incident wavefront.

i = angle of incidence
r = angle of reflection

Refraction

Figure 3.8 shows that when the waves enter the shallow water above the glass plate, their speed is reduced. Their frequency stays the same so their wavelength is also reduced. The refracted waves change direction.

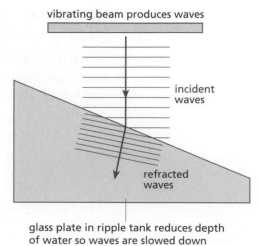

Figure 3.8 Refraction of waves in a ripple tank

● **Refraction explained by wave theory**

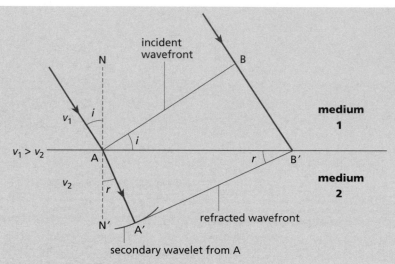

Figure 3.9 Refraction of a straight wavefront

Speed is reduced in the second medium. The distance travelled by the secondary wavelet AA' is therefore less than BB', the distance travelled by the incident wavelet. So the refracted wavefront is *not* at the same angle as the incident wavefront.

Diffraction

● **Diffraction from a narrow gap**

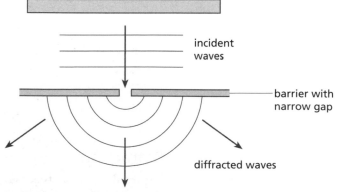

Figure 3.10 Diffraction of waves in a ripple tank by a gap that is narrow compared with the wavelength

Speed, wavelength and frequency are unchanged by diffraction.

Common errors

✗ Carelessness in drawing diffraction diagrams.
✗ Poor semi-circles, not drawn with the middle of the narrow gap as their centre.
✗ Different wavelengths of incident and diffracted waves.
✗ Varying wavelength of diffracted waves.
✓ To draw the diffracted waves accurately, measure the wavelength of the incident waves. This is the radius of the first circle. For each subsequent circle, the increase of radius must be measured to be the same as the wavelength.
✓ Careful, accurate measuring and drawing is *essential* to produce good diagrams. ■

● **Diffraction from a wide gap**

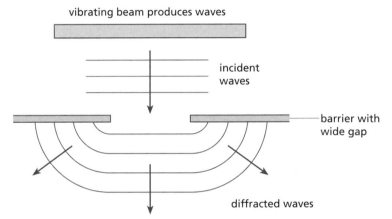

vibrating beam produces waves

incident waves

barrier with wide gap

diffracted waves

Figure 3.11 Water waves diffracting after passing through a gap of width greater than their wavelength

Examiner's tip
▶ The centres of the quarter-circles must be at the edges of the wide gap.

● **Diffraction explained by wave theory**

- Narrow gap: when the gap is one wavelength wide or less, the gap acts as the only centre of a secondary wavelet, so circular waves spread out from the gap.
- Wide gap: the wavelets from the edges of the gap spread out to produce waves of quarter-circle shape, while in front of the gap itself there is a straight line joining the wavelets, parallel to the gap.

Examiner's tips
▶ You need to be able to give the meaning of the term 'wavefront'.
▶ You need to be able to explain reflection, refraction and diffraction using wave theory of secondary wavelets.

Light

Light moves as waves of very small wavelength but it is often convenient to use light rays to work out and explain the behaviour of light.

- A light ray is the direction in which light is travelling and is shown as a line in a diagram.
- An object is what is originally observed.

• An optical image is a likeness of the object, which need not be an exact copy.
• A real image is formed on a screen.
• A virtual image is observed when rays appear to come from it.

Reflection

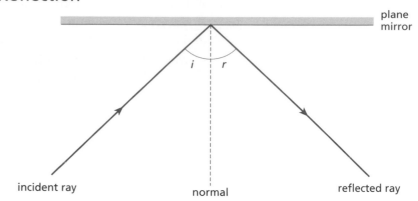

Figure 3.12 Reflection of light by a plane mirror

i = angle of incidence = angle between **normal** and incident ray

r = angle of reflection = angle between **normal** and reflected ray

angle of incidence, i = angle of reflection, r

Sample question Draw a diagram to show the path of a ray striking a plane mirror with an angle of incidence of 35°. Mark and label the incident ray, normal, reflected ray and angle of reflection. [4 marks]

Student's answer The student's answer is shown by dashed lines in Figure 3.13. [2 marks]

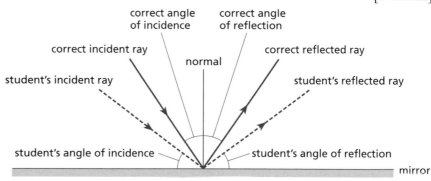

Figure 3.13

Examiner's comments *The student has measured the angle of incidence away from the mirror line, not away from the normal. The normal is correct, as is the reflected ray for the incident ray drawn. The angle of reflection is also incorrectly measured away from the mirror line.*

Correct answer The correct answer is shown by solid lines in Figure 3.13. [4 marks]

Common error

✗ Finding the angles of incidence and reflection by measuring between the ray and the mirror. ■

Formation of a virtual image by a plane mirror

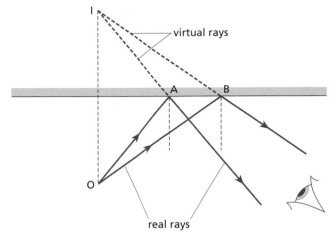

Figure 3.14 Construction to find the image in a plane mirror

The image of the object O is not formed on a screen. It is at I, where the rays appear to come from. The properties of an image in a plane mirror are:

1) it is virtual
2) it is the same size as the object
3) the line joining the object and the image is perpendicular to the mirror
4) it is the same distance behind the mirror as the object is in front
5) it is laterally inverted.

> **Examiner's tips**
> ▶ You must be able to draw simple constructions.
> ▶ Hints for drawing a construction to show the position of the image of a point object in a plane mirror:
>   **1)** Carefully measure the distance of the object from the mirror.
>   **2)** Mark the image the same distance behind the mirror as the object is in front. The object and image should be on a line at right angles to the mirror line (OI in Figure 3.14).
>   **3)** Draw two lines from the image towards the eye; draw dotted lines behind the mirror where they represent virtual rays.
>   **4)** Join up two lines from the object to where the previous two lines cut the mirror line (A and B in Figure 3.14).
>   **5)** Mark arrows on the real rays and label the diagram as necessary.

Refraction

When a ray is travelling at an angle to a surface and enters a material where it travels slower, it changes direction *towards* the normal. This is called **refraction**. (This was explained earlier by wave theory.) When a ray leaves this material, it is refracted *away from* the normal.

> **Examiner's tips**
> ▶ You must be able to describe an experiment to demonstrate the refraction of light.
> ▶ You should do this experiment in a darkened room. Direct a narrow beam of light at an angle to the side of a glass block placed on a large piece of paper. Mark on the paper the paths of the beams entering and leaving the block. By joining up the lines after removing the block, you can draw the path of the light as it travelled through the block. Figure 3.15 shows the paths of the rays in this experiment.

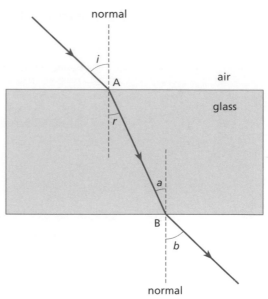

Figure 3.15 Refraction of a ray through a glass block

For refraction at A, where the ray enters the glass:

i = angle of incidence = angle between *normal* and incident ray

r = angle of refraction = angle between *normal* and refracted ray

The ray is refracted towards the normal so the angle of refraction is less than the angle of incidence.

For refraction at B, where the ray leaves the glass:

a = angle of incidence = angle between *normal* and incident ray

b = angle of refraction = angle between *normal* and refracted ray

The ray is refracted away from the normal so the angle of refraction is greater than the angle of incidence. When the block is parallel-sided, the ray leaving is parallel to the ray entering.

Common error

✗ Finding the angles of incidence and refraction by measuring between the ray and the surface of the new material. ◼

Sample question Copy and complete Figure 3.16a to show the path of the ray through the glass prism as it is refracted twice. Show *both* normals.

[4 marks]

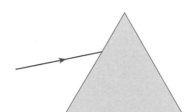

Figure 3.16a

Student's answer   The student's answer is shown by dashed lines in Figure 3.16b.

[3 marks]

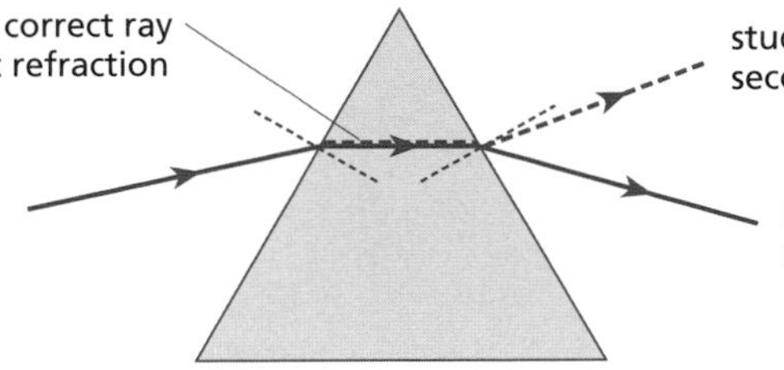

student's correct ray after first refraction

student's ray after second refraction

correct ray after second refraction

Figure 3.16b

Examiner's comments   *The student has correctly drawn the refracted ray within the prism and both normals. However, the student's second refraction, as the ray left the prism, was towards the normal. When a ray moves into a region where it travels faster, it is refracted away from the normal.*

Correct answer   The correct rays are shown by solid lines in Figure 3.16b.

[4 marks].

The amount of refraction is determined by the **refractive index** – the ratio of speed of light in air to speed of light in glass.

$$\text{Refractive index, } n = \frac{\text{speed of light in air}}{\text{speed of light in glass}}$$

Examiner's tip
▶ You must be able to recall and use the equation: $\frac{\sin i}{\sin r} = n$

Sample question   Light travels at 3×10^8 m/s in air and 2.25×10^8 m/s in water. Calculate

a) the refractive index, n, of water,
b) the angle of refraction for a ray approaching water with an angle of incidence of 55°.   [4 marks]

Student's answer   **a)** $n = \dfrac{3 \times 10^8}{2.25 \times 10^8} = 1.33$ [2 marks]

b) $r = 34°$ [0 marks]

Examiner's comments   **a)** *Correct answer with working.*
b) *The answer is only slightly inaccurate but there is no working, which means the examiner has no way of knowing whether the student made a small mistake or was completely wrong and simply close to the correct answer by luck.*

Correct answer   **a)** $n = \dfrac{3 \times 10^8}{2.25 \times 10^8} = 1.33$ [2 marks]

b) $\sin r = \dfrac{\sin i}{1.33} = \dfrac{0.819}{1.33} = 0.616$

$r = 38.0°$ [2 marks]

● **Try this** The answers are given on **p. 103**.

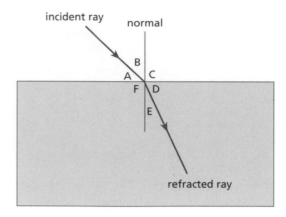

Figure 3.17

3 Figure 3.17 shows a light ray entering a glass block of refractive index 1.5. The angle of incidence is changed so that angle A is now 60°.
   a) Which of angles A–F is the angle of incidence?
   b) Which of angles A–F is the angle of refraction?
   c) Work out the values of angles B–F.

Internal reflection and critical angle

When a ray inside a block of glass or tank of water passes out into the air, some light is reflected internally as well as being refracted away from the normal.

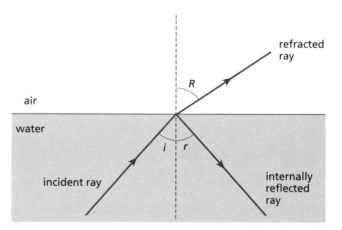

Figure 3.18 Rays at a water–air boundary

i = angle of incidence = angle between *normal* and incident ray

r = angle of internal reflection = angle between *normal* and internally reflected ray

R = angle of refraction = angle between *normal* and refracted ray

The law of reflection still applies, so $i = r$.

The greater the angle of incidence, the more energy goes into the internally reflected ray, which becomes brighter. The greatest angle of incidence when refraction can still occur is called the **critical angle**; in this case (shown in Figure 3.19) the angle of refraction is 90° and the refracted ray travels along the surface.

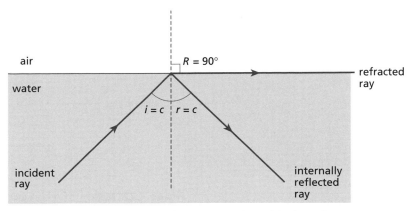

Figure 3.19 Angle of incidence is the same as the critical angle

angle of incidence, $i = c$, critical angle

angle of refraction, $R = 90°$

If the angle of incidence is greater than the critical angle, there is no refracted ray and all the energy is in the bright internally reflected ray. This is called **total internal reflection** (Figure 3.20).

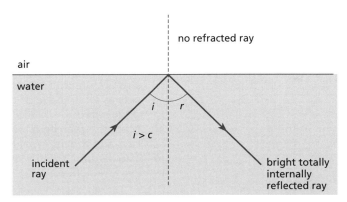

Figure 3.20 Total internal reflection

Water has a refractive index of 1.33 and a critical angle of 48.5°. The following calculation confirms that there can be no refracted ray with an angle of incidence of 50°.

For light *leaving* water:

$$n = \frac{1}{1.33} = 0.75$$

$$i = 50°, \sin i = 0.766$$

$$\sin r = \frac{\sin i}{n} = \frac{0.766}{0.75} = 1.02$$

1.02 is a non-existent value for the sine of an angle.

● **Optical fibres**

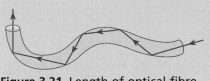

Figure 3.21 Length of optical fibre

Each time the light strikes the wall of the optical fibre, the angle of incidence is greater than the critical angle and so total internal reflection occurs. There is very little loss of energy. The light can be considered 'trapped' in the optical fibre and can travel long distances, even if the fibre is bent in order to carry information or illuminate and view inaccessible places.

Formation of a real image by a converging lens

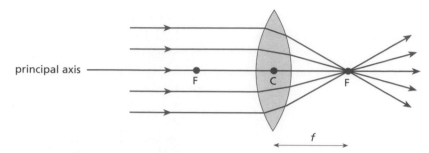

Figure 3.22 Action of a converging lens on a parallel beam of light

All rays of light parallel to the principal axis are refracted by the lens to pass through the **principal focus**, F. The distance between F and the optical centre, C, is called the focal length, f.

Examiner's tip

▶ You must be able to draw ray diagrams to illustrate the formation of an image by a converging lens. Follow these steps:
1) Preliminary – draw the principal axis, the object, a vertical line for the lens and mark the principal focus, F.
2) Ray 1 – draw a ray from the top of the object to the line of the lens, parallel to the principal axis, and continue this ray to pass through F and a few centimetres beyond. This is because all rays parallel to the principal axis pass through the principal focus.
3) Ray 2 – draw this ray from the top of the object through C to pass straight on until it cuts Ray 1. This is because the centre of the lens acts as a thin pane of glass so rays pass through the centre undeviated.
4) Mark the real, inverted image between the intersection of the rays and the principal axis.
5) The diagram should look like Figure 3.23.

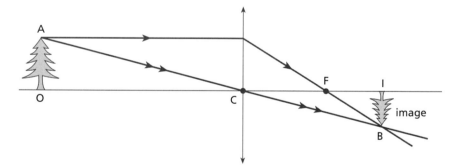

Figure 3.23 Ray diagram for a converging lens

If the object is placed more than one focal length behind the lens, the image will always be real and inverted. Depending on where the object is placed, the image may be magnified, the same size as the object or diminished.

Common error

✗ Inaccurate drawing work leading to very poor answers.
✓ Because rays intersect at small angles, errors can easily be magnified. It is essential to work carefully with a sharpened pencil; distances should be measured to within 0.5 mm or less and lines drawn *exactly* though points. ■

● **Try this**   The answers are given on **p. 103**.

4 An object of height 1.5 cm is placed 4.5 cm from a thin converging lens
of focal length 3 cm. Draw a ray diagram to find **a)** the nature, **b)** the
size, **c)** the position of the image.

Examiner's tips

▶ Draw a third ray for accuracy and to check for any major mistakes.
▶ There is in fact a principal focus on each side of the lens. All rays
passing through the second principal focus and then reaching the lens
are refracted parallel to the principal axis. Add to your diagram for
Try this question 4 as follows:
1) Mark the second principal focus F on the object side of the lens.
2) Draw a line from the top of the object through this new principal
focus and extend it to the lens.
3) From where the ray strikes the lens, it continues parallel to the
principal axis until it meets the other two rays.
4) The three rays theoretically meet at a point. In practice, due to
human error, even with very careful work the rays may meet as a
small triangle.
5) If they do not meet at a point or as a small triangle, go back and
check that each ray is correctly and accurately drawn.

Formation of a virtual image by a converging lens

If the object is placed closer to the lens than the principal focus,
the rays leaving the lens do not converge to form a real image.

● **Try this**   The answers are given on **p. 103**.

5 An object of height 1.5 cm is placed 2 cm from a thin converging lens of
focal length 3 cm. Draw a ray diagram to show two rays from the object
passing through the lens.
a) What can you say about the two rays as they leave the lens?
b) With dotted lines, draw your rays back behind the object. Find
i) the nature, **ii)** the size, **iii)** the position of the image.

● **Magnifying glass**

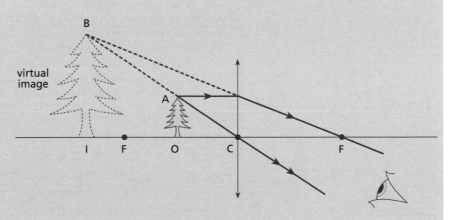

Figure 3.24 Ray diagram for a converging lens

Figure 3.24 shows how a converging lens can be used as a
magnifying glass. The object is placed less than one focal length
behind the lens. No real image is formed but the eye sees the rays
diverging from the magnified virtual image.

Dispersion of light

White light is made up of seven colours. Each colour is refracted by a different amount in glass. If a beam of white light falls on a glass prism, it is dispersed into a **spectrum** of the seven colours.

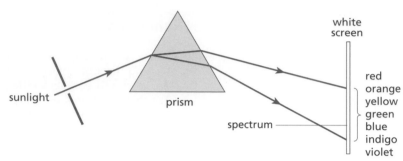

Figure 3.25 Forming a spectrum with a prism

Electromagnetic spectrum

All types of waves that make up the electromagnetic spectrum have properties in common:

1) They can travel through a vacuum at the same high speed, which is much faster than other types of waves that travel through a material.

> **Examiner's tip**
> ▶ You must be able to recall the speed of electromagnetic waves in a vacuum. It is 3×10^8 m/s.

2) They show the normal wave properties of reflection, refraction and diffraction.
3) They are transverse waves.
4) They travel due to moving electric and magnetic fields.

> **Examiner's tip**
> ▶ Remember that all electromagnetic waves have the same speed in a vacuum and the wave equation $v = f\lambda$ applies, so the higher the wavelength the lower the frequency.

The Sun and other stars give off a wide range of types of electromagnetic waves, which travel through space to Earth. Much of this radiation is stopped by Earth's atmosphere and can only be detected by satellites in orbit outside the atmosphere.

The types of electromagnetic waves in order of increasing wavelength are described below.

- **Gamma rays** are produced by radioactive substances. They are very dangerous to living matter. They are used to kill cancer cells and dangerous bacteria.
- **X-rays** are produced in high voltage X-ray tubes. They can also be dangerous to living matter. They are absorbed differently by different types of matter so can produce shadow pictures of inside the human body or inaccessible metal structures.

• **Ultraviolet (UV) radiation** is produced by the Sun, special UV tubes and welding arcs. The radiation can cause sunburn and skin cancer; it also produces vitamins in the skin and causes certain substances to fluoresce. This fluorescence can reveal markings that are invisible in light. Special photographic films react to UV radiation.

Common error

✗ Using the expression 'ultraviolet light'.
✓ The correct expression is 'ultraviolet radiation' – UV radiation is not part of the visible spectrum so must *not* be called 'light'. This misconception might occur because normally UV lamps give off blue and violet light as well as UV radiation. ■

• **Visible light** is a very narrow range of wavelengths that can be seen by the human eye as the colours of the visible spectrum from violet to red. Normal photographic films are designed to be sensitive to visible light.

Examiner's tip
▶ You must be able to use the term 'monochromatic'. It literally means 'all of the same colour' so when applied to light it means 'all of the same wavelength'.

• **Infrared (IR) radiation** is produced by hot objects and transfers heat to cooler objects. Hot objects below about 500 °C produce only IR radiation; above this temperature, visible light is also radiated. Special photographic films react to IR radiation. Night-vision goggles detect the IR radiation given off by warm objects.
• **Radio waves** come in a wide range of wavelengths that can be split into these groups:
   1) Microwaves have very short wavelengths, close to that of IR radiation. They are used for telecommunication, both ground-based and via satellites. They are also used for radar and cooking.
   2) VHF (very high frequency) and UHF (ultra high frequency), of longer wavelengths, are used for television and radio.
   3) Traditional radio waves (short, medium and long wave) have the longest wavelengths.

Sample question
a) Which of the following types of waves is *not* an electromagnetic wave: light, X-rays, radio, sound? [1 mark]
b) State one property of this wave that it has in common with electromagnetic waves. [1 mark]
c) State one property of this wave that is different from electromagnetic waves. [1 mark]

Student's answer
a) sound [1 mark]
b) wavelength [0 marks]
c) reflection [0 marks]

51

Examiner's comments	**a)** *Correct answer*	
	b) *Incorrect answer – although both types of waves do have a wavelength, its value can vary over a wide range so it is unlikely to be the same for each type of wave*	
	c) *Incorrect answer*	

Correct answer	**a)** sound	[1 mark]
	b) both can be reflected (or refracted or diffracted)	[1 mark]
	c) sound waves are longitudinal	[1 mark]

● **Try this**   The answers are given on **p. 103**.

6 An observatory receives X-rays and gamma rays from a star.
   a) Which type of radiation has the higher wavelength?
   b) Which type of radiation has the higher frequency?
   c) Light from the star takes four years to reach the observatory. Do the X-rays take less, more or the same time to reach the observatory?
   d) Work out the distance from the observatory to the star in kilometres.

Sound

- Sound waves are **longitudinal** waves that are produced by a vibrating source, which causes a material to vibrate. A material or medium is required to transmit sound waves.
- Although normally observed in air, sound waves can travel through liquids and solids, e.g. sea creatures communicate by sound waves travelling through water.
- As sound travels through a material, compressions and rarefactions occur (see Figure 3.2). Compressions are regions where particles of material are closer together. Regions of material are rarefied where the particles move further apart (at rarefactions).
- The human ear can hear sound in air in the frequency range of 50 Hz to 20 000 Hz (20 kHz). This is called the **audible range**.
- The greater the amplitude of sound waves, the louder the sound.
- The greater the frequency of sound waves, the higher the pitch.
- Sound waves can be reflected, especially from large, hard, flat surfaces. The reflected sound is called an **echo**.

Examiner's tips
▶ You must be able to describe an experiment to find the speed of sound in air. (See the **Sample question** below.)
▶ The speed of sound in air at normal temperatures is about 340 m/s. Sound travels about four times faster in water at about 1400 m/s. The speed of sound in solids is very high and in metals is of the order of 10–15 times faster than in air. Supplement students should know these values approximately.

Sample question   A student stands 100 m from a large building and claps her hands regularly at a rate of 16 claps every 10 s. She hears each clap coincide exactly with the echo from the clap before. Work out the speed of sound in air.   [4 marks]

Student's answer	Time between claps $= \frac{16}{10} = 1.6\,$s	[0 marks]
	Distance travelled $= 2 \times 100 = 200\,$m	[1 mark]
	Speed $= \frac{200}{1.6} = 125\,$m/s	[1 mark]

Examiner's comments *The student calculated the number of claps per second instead of the time from one clap to the next. No further error was made in calculating the speed so the student gained the last two marks.*

Correct answer	Time between claps $= \frac{10}{16} = 0.625\,$s	[2 marks]
	Distance travelled $= 2 \times 100 = 200\,$m	[1 mark]
	Speed $= \frac{200}{0.625} = 320\,$m/s	[1 mark]

Common errors

✘ In problems involving echoes, taking the distance from the observer to the reflecting surface as the distance travelled by the sound.

✔ When the source and observer are at the same place, the sound travels *twice* the distance between the observer and the reflecting surface.

✘ Doubling the distance when no echo or reflection is involved. ◼

Sample question A railway worker gives a length of rail a test blow with a hammer, striking the end of the rail in the direction of its length. A sound of frequency 10 kHz travels along the rail. Use your knowledge of the order of magnitude of the speed of sound in a solid to estimate the wavelength of the sound. [4 marks]

Student's answer	Estimated speed of sound $= 3000\,$m/s	[2 marks]
	Wavelength, $\lambda = \frac{v}{f} = \frac{3000}{10\,000} = 0.3\,$m	[2 marks]

Examiner's comment *The estimate for speed of sound is on the low side as sound travels faster in steel than in less rigid solids. Full marks are still, just, awarded as only an order of magnitude was asked for. The rest of the answer is correct.*

Correct answer	Estimated speed of sound $= 5000\,$m/s	[2 marks]
	Wavelength, $\lambda = \frac{v}{f} = \frac{5000}{10\,000} = 0.5\,$m	[2 marks]

● **Try this** The answer is given on **p. 103**.

7 A research ship uses an echo-sounder to locate a shoal of fish. It receives back an echo 37.1 ms after a sound is transmitted. Work out the depth of the shoal below the ship. (Speed of sound in water = 1400 m/s.)

N.B. Extension students are expected to know the order of magnitude of the speed of sound in water.

TOPIC 4 Electricity and Magnetism

Key objectives

- To be able to state the properties of magnets
- To be able to describe the magnetic field due to electric currents
- To be able to state the nature of charge, current, voltage and resistance
- To be able to draw and interpret basic circuit diagrams and solve simple circuit problems
- To be able to describe the action of logic gates on their own and in circuits
- To be able to state the basic hazards of electricity
- To know how a magnetic field can cause a force on a current-carrying conductor and relate this force to a d.c. electric motor
- To know the principle of electromagnetic induction and how it is applied in an a.c. generator and a transformer
- To be able to describe the production, deflection and detection of charged particles

Key definitions

Term	Definition
d.c.	Direct current which always flows in one direction
a.c.	Alternating current which repeatedly reverses its direction of flow
Magnetic field	A region where a magnet experiences a force
Electric charge	An excess or lack of electrons measured in coulombs
Current	The rate of flow of charge measured in amps
emf	Electromotive force: the source of electrical energy in a circuit measured in volts
p.d.	Potential difference: the voltage between two points in a circuit
Resistance	Resistance = $\dfrac{\text{p.d.}}{\text{current}}$ measured in ohms
LDR	The resistance of a light dependent resistor (LDR) depends on the intensity of light falling on it
Thermistor	The resistance of a thermistor depends on its temperature
Cathode ray	A beam of electrons flowing in a vacuum
Analogue	In an analogue circuit, the voltages at various point can vary continuously to take any value in the available range
Digital	In a digital circuit, the voltage at various points can only be in one of two states – either zero or maximum voltage

Key ideas   Magnetism

- A magnetic field is a region where a magnetic material experiences a force.
- Magnetic materials are chiefly the ferrous metals, iron and steel, and their alloys. Cobalt and nickel are also magnetic materials.
- A magnetic field can be produced by a permanent magnet or a wire carrying an electric current. A magnetic field also exists around the Earth due to convection currents in its molten iron core.

> **Examiner's tips**
> ▶ You must be able to describe an experiment to identify the pattern of a magnetic field around a bar magnet.
> ▶ This experiment can be done either with a plotting compass or iron filings.

● **Plotting compass method**

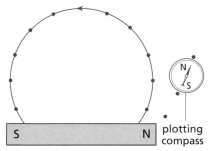

Figure 4.1 Plotting field lines

The magnet is placed on a sheet of paper and a small plotting compass placed near one pole. Dots are marked on the paper at the positions of the ends of the compass needle. The compass is moved so the end that was over the first dot is now over the second dot. The other end is marked on the paper as a third dot. This process is continued until the opposite pole of the magnet is reached. Joining the dots with a smooth line shows the field, with the direction being given by the compass arrow. This can be repeated for further lines starting at different points.

● **Iron filings method**

Iron filings are sprinkled on a piece of paper placed over a magnet. When the paper is tapped gently, then the filings will be seen to line up with the field lines. The field pattern can then be drawn along these lines on the paper.

Sample question

Describe how to plot the magnetic field around a bar magnet using iron filings. [4 marks]

Student's answer

The iron filings should be spread around the magnet and the pattern drawn. [1 mark]

Examiner's comments

The student basically knows the right experiment but the description is very vague and lacking in essential detail so few marks are awarded.

Correct answer

Place a piece of paper on top of a bar magnet. [1 mark] Sprinkle iron filings thinly and evenly over the paper. [1 mark] Give the paper a gentle tap. [1 mark] Draw the field pattern on the paper along the lines of the filings. [1 mark]

- One end of a magnet suspended in the Earth's magnetic field will swing round to point towards the Earth's north pole. This is called the north or N pole of the magnet because it is 'north seeking'. The other end of the magnet is the south or S pole.
- If two magnets are close together, N and N or S and S will repel but N and S will attract.

Examiner's tips
▶ Unlike poles attract.
▶ Like poles repel.
▶ The Earth's north pole attracts the north pole of a magnet so magnetically it is a south pole.

● **Try this**   The answers are given on **p. 103**.

1 A bar magnet is suspended from its mid-point by a thread.
   a) Which pole will swing towards the Earth's north pole?
   Another bar magnet is brought close to the pole of the suspended magnet that is pointing away from the Earth's north pole and does *not* cause the suspended magnet to swing round.
   b) Which pole of the new magnet is closest to the suspended magnet?

- If a piece of iron or steel is placed in a magnetic field it too will become magnetised. This is called induced magnetism.
- Steel is hard to magnetise by induction but once magnetised keeps its magnetism. Materials that are hard to magnetise and demagnetise are called **magnetically hard** materials. Permanent magnets, which need to maintain a high level of magnetism for a long time, are made from these materials.
- The uses of permanent magnets include loudspeakers, electrical meters and small electric motors, as used in domestic appliances.
- Generally, iron is easily magnetised but loses its induced magnetism when removed from the magnetic field. Materials that easily magnetise and demagnetise are called **magnetically soft** materials.
- Magnets that need to be switched on and off readily or have their fields reversed are made from soft iron. Their uses include electromagnets, large scale motors and generators, relays and transformers.

Common misconception

✗ Magnets attract all metals.
✓ Magnets only attract cobalt, nickel and ferrous metals. ■

- Materials can be magnetised by placing them in a strong magnetic field. This is best done in the field produced inside a long coil or solenoid with a high direct current (d.c.) flowing through it.

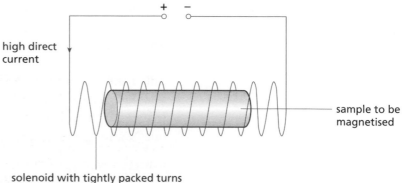

Figure 4.2 Magnetising in a solenoid

- An old method of magnetising a steel bar is to 'stroke' a magnet along the length of the bar repeatedly in the same direction.

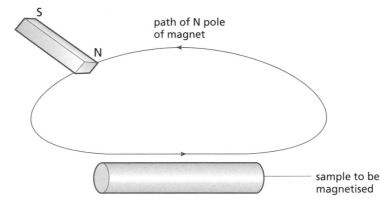

Figure 4.3 Magnetising by stroking

- Simple methods of demagnetising a material are by heating until red hot or striking it a sharp blow with a hammer.
- A material can also be demagnetised electrically. The magnet is placed in a solenoid supplied with an alternating current (a.c.). The current is gradually reduced to zero. During this process, the material is magnetised in opposite directions every half cycle of the a.c., less and less strongly until no magnetism remains.

Sample question One method of demagnetising a bar magnet is to place it in a solenoid with a high alternating current and slowly withdraw the magnet from the solenoid. Explain how this demagnetises the magnet. [4 marks]

Student's answer The magnet gets tired of the alternating field in the solenoid and loses magnetism. [2 marks]

Examiner's comments *The student has correctly recognised that there is an alternating field inside the solenoid but fails to understand the principle of magnetisation.*

Correct answer The alternating current causes the direction of the field in the solenoid to alternate. [1 mark] When inside the solenoid, the magnet is repeatedly magnetised in alternating directions by the strong alternating field. [1 mark] Outside the solenoid, the field weakens with distance moved away. As each part of the magnet is withdrawn, it is magnetised in alternating directions less and less strongly [1 mark] until no magnetism remains. [1 mark]

● **Try this** The answers are given on **p. 103**.

2 Non-magnetised bars of hard steel, soft iron and aluminium are placed in a long coil carrying a very high d.c. current. On removal the bars are tested by seeing how many paper clips they can pick up. Which bar picks up **a)** no paper clips, **b)** a few paper clips, **c)** a lot of paper clips?

Electromagnetism

- A wire or coil carrying an electric current produces a magnetic field.
- The higher the current, the stronger the magnetic field.
- If the current reverses, the direction of the magnetic field is also reversed.

Field due to a straight wire

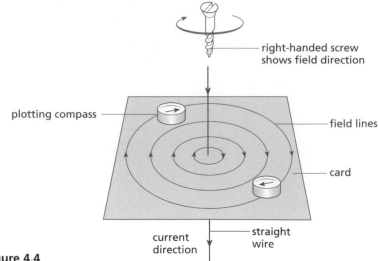

Figure 4.4

- In a field due to a straight wire, the further a point is from the wire, the weaker the magnetic field.

Field due to a solenoid

A solenoid is a long cylindrical coil. When a current flows, the field pattern outside the solenoid is similar to that of a bar magnet. Inside the solenoid, there is a strong field parallel to the axis. The **right-hand grip rule** (Figure 4.5b) gives the direction of the field. The fingers of the right hand are curled as if to grip the solenoid, pointing in the direction of the current, and the thumb points to the N pole.

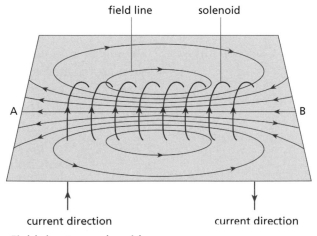

Figure 4.5a Field due to a solenoid

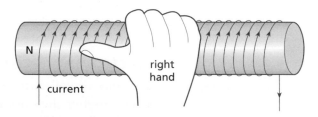

Figure 4.5b The right-hand grip rule

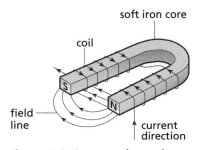

Figure 4.6 C-core or horseshoe electromagnet

The closeness of the field lines in Figure 4.5a indicate the strength of the field. The field is very strong within the solenoid. Outside the solenoid, the further away a point is, the weaker the field.

The strength of the magnetic field of a solenoid is greatly increased if there is a core of soft iron inside the coils. As the field can be switched on and off with the current, this is the basis for the **electromagnet** (Figure 4.6).

Uses of electromagnets

Sample question Figure 4.7 shows an electric bell.

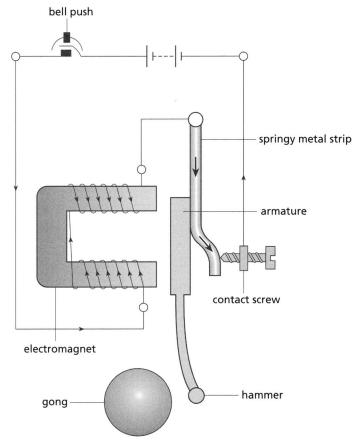

Figure 4.7

When the bell push is pressed, a current flows through the electromagnet. The armature is attracted to the electromagnet and the hammer strikes the gong. The movement of the armature breaks the circuit that supplies current to the electromagnet. This releases the armature, which springs back. The circuit is re-made, the process repeats and the bell rings continuously, for as long as the bell push is pressed. Choose a suitable material for **a)** the armature, **b)** the core of the electromagnet. Give your reasons. [4 marks]

Student's answer **a)** Use hard steel, as a strong magnet is required. [0 marks]
b) Use soft iron, as the electromagnet must be switched on and off repeatedly. [2 marks]

Examiner's comments **a)** *The armature is not a permanent magnet but becomes an induced magnet only when the electromagnet produces a field. When no current flows, the armature needs to spring back so there must be no permanent attraction to the core of the electromagnet. Therefore soft iron should be used, not hard steel.*
b) *Correct answer*

Correct answer
a) Use soft iron, as the armature must only be attracted to the electromagnet when the current is switched *on*.　[2 marks]
b) Use soft iron, as the electromagnet must be switched on and off repeatedly.　[2 marks]

> **Examiner's tips**
> ▶ You need to be able to describe applications of the magnetic effects of a current, including the action of a relay (described below).
> ▶ You need to be able to show understanding of the use of a relay in switching circuits.
> ▶ You need to be able to describe the operation of an electromagnet, an electric bell and a moving-coil loudspeaker.

● **Try this**　The answers are given on **p. 103**.

3 A loudspeaker is made up essentially of a stationary magnet that is close to a small coil fixed to a paper cone. The signal from the amplifier is a small alternating current supplied to the coil. Describe briefly **a)** the variation of magnetic field produced by the coil, **b)** the variation of the magnetic force on the coil, **c)** the motion of the paper cone.

● **Electromagnetic relay**　A relay is a device which enables one electric circuit to control another. It is usually used when the first circuit only carries a small current, e.g. in an electronic circuit, and the second circuit requires a much higher current.

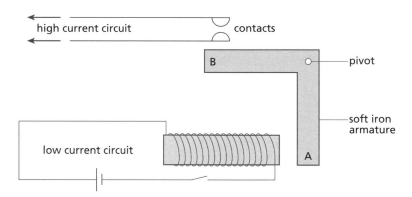

Figure 4.8 An electromagnetic relay

When the switch is closed in the low current circuit, current flows to the electromagnet, which attracts end A of the soft iron armature. The armature pivots and end B moves up to close the contacts in the high current circuit. This circuit is now complete and the high current flows through the device, e.g. a motor, a heater or an alarm bell.

Electric charge
- An **electron** is a subatomic particle with a negative electric charge. Certain materials can lose or gain electrons.
- Atoms, and objects composed of them, are normally electrically neutral. The number of electrons (negative charges) balances the positive atomic charges. If an object gains extra electrons, it becomes negatively charged. If it loses electrons, it becomes positively charged.

- The charge on an electron is the smallest possible quantity of charge. Charge is measured in coulombs (C) and the charge on one electron is 1.6×10^{-19} coulombs.
- If two charged objects are close together, + and + or − and − will repel but + and − will attract.

Common misconception

✗ There is only a force of attraction when both objects are charged.
✓ Charged objects attract uncharged objects as well as charged objects. ▮

- When certain materials, e.g. polythene, are rubbed with a cloth, electrons are moved from the cloth on to the polythene. The polythene gains electrons and becomes negatively charged.
- When certain other materials, e.g. cellulose acetate, are rubbed with a cloth, electrons are moved from the cellulose acetate on to the cloth. The cellulose acetate loses electrons and becomes positively charged.
- The fact that an object is electrically charged can be detected as shown in Figure 4.9.
- Electrical conductors are materials in which electrons can move freely from atom to atom so electric charge can flow readily. All metals and some forms of carbon are good conductors.
- Electrical insulators are materials in which electrons are firmly held in their atoms and cannot move so electric charge cannot flow. Most plastics are good insulators.

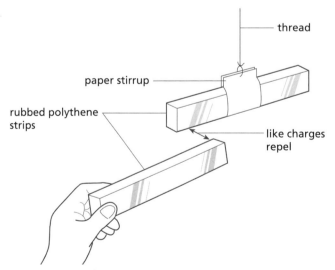

Figure 4.9 Investigating charges

Two polythene rods are rubbed with a cloth to become negatively charged. One rod is suspended freely on a thread. The other rod is brought close to one end of the suspended rod, which rotates away due to repulsion between like charges.

Similarly, if a charged cellulose acetate rod is brought close to one end of the suspended, charged polythene rod, the latter will rotate closer due to attraction between unlike charges.

• Another test for electric charge is to use the **gold-leaf electroscope**, as shown in Figure 4.10.

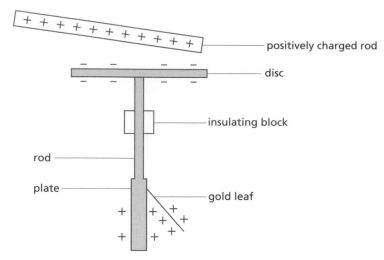

Figure 4.10 Gold-leaf electroscope

The gold–leaf electroscope is composed of a metal rod held inside a glass case by a block of insulating material. A metal disc is connected to the top of the rod and a metal plate to the bottom, with a thin leaf of gold foil attached. Figure 4.10 shows the electroscope when a positively charged rod is brought close to the disc. Free electrons in the metal move up to the disc due to attraction of the positively charged rod. This leaves the plate and the gold leaf positively charged. There is a repulsive force between them and the gold leaf is deflected.

Sample question An inkjet printer produces a stream of very small droplets from a nozzle. The droplets are given a negative electrostatic charge and then pass between two plates with positive and negative charge as shown in Figure 4.11a.

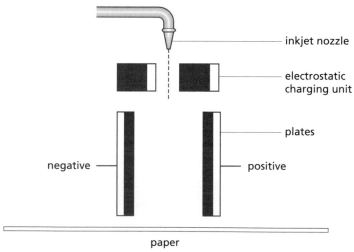

Figure 4.11a

a) State the type of field in the region between the charged plates.
[1 mark]

b) State and explain the force acting on the droplets in this region. [2 marks]

c) Extend the dashed line to complete the path of the droplets.
[2 marks] [3 marks]

Student's answer
a) There is an electric field in this region.   [1 mark]
b) There is a force to the right on the ink droplets.   [1 mark]
c) The student's answer is shown by the dashed extension in Figure 4.11b.   [2 marks]

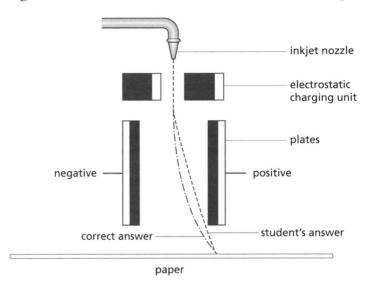

inkjet nozzle

electrostatic charging unit

plates

negative

positive

correct answer ——————— student's answer

paper

Figure 4.11b

Examiner's comments
a) *Correct answer*
b) *The student correctly stated that the force is to the right but failed to explain it.*
c) *Acceptable answer for Core students.*
   The path between the plates must be curved, as the droplets are accelerated to the right by the electric force.

Correct answer
a) There is an electric field in this region.   [1 mark]
b) There is a force to the right on the ink droplets, which are attracted to the positive plate.   [2 marks]
c) The dashed extension in Figure 4.11b is an acceptable answer for Core students.   [2 marks]
   The correct answer is shown by the chain-dotted extension in Figure 4.11b.   [3 marks]

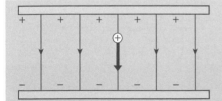

Figure 4.12 Uniform field between parallel conducting plates with opposite charge

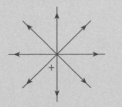

Figure 4.13 Radial field around a positive point charge

• An **electric field** is a region where an electric charge experiences a force.

• The field can be represented by lines of force in the direction that a positive charge would move if placed at a point in the field (Figure 4.12 and Figure 4.13).

A conducting object can be charged by induction as shown in the example in Figure 4.14 overleaf, with two conducting spheres. The neutral, uncharged metal spheres are in contact and a charged rod is brought close. This repels electrons, charging the far side of sphere B negatively and the near side of sphere A positively (Figure 4.14a). Keeping the charged rod close to sphere A, sphere B is moved away (Figure 4.14b). When the charged strip is removed, the spheres have been charged (Figure 4.14c).

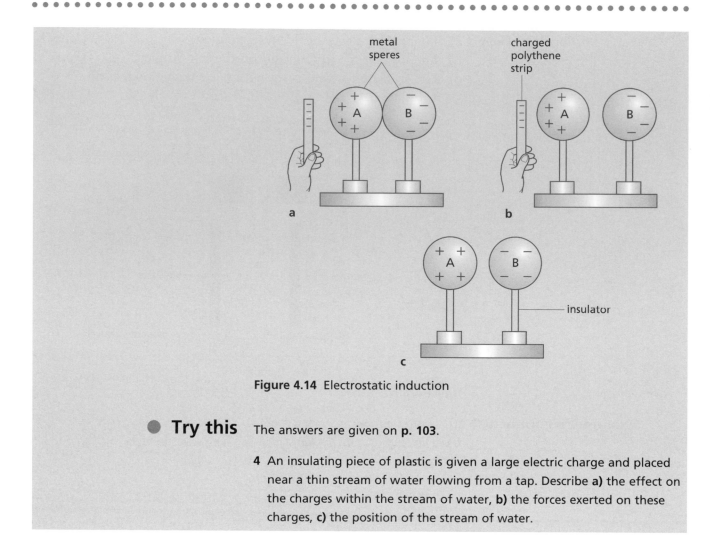

Figure 4.14 Electrostatic induction

● **Try this**   The answers are given on **p. 103**.

4 An insulating piece of plastic is given a large electric charge and placed near a thin stream of water flowing from a tap. Describe **a)** the effect on the charges within the stream of water, **b)** the forces exerted on these charges, **c)** the position of the stream of water.

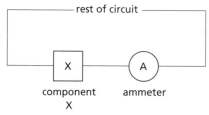

Figure 4.15 Measuring the current through component X

Electric current

- Electric current is a flow of charge and is measured in **amperes**, usually abbreviated to amps (A).
- Electric current flows from the + terminal of a battery or power supply through the circuit to the − terminal.
- Electric current is measured by an ammeter, which must be connected **in series**.
- Figure 4.15 shows an ammeter in series with component X.

Common error

✓ Current does not flow. It is charge that flows thus creating a current *in* something, not *through* it.

✗ Putting an ammeter **in parallel** with the part of the circuit where current is to be measured.

✓ The current to be measured must flow *through* the ammeter, which needs to be connected **in series**. ■

• One ampere is a flow of one **coulomb** (1 C) of charge in one second.

> **Examiner's tip**
> ▶ You must be able to recall and use the equation: $I = \frac{Q}{t}$
>
> I = current, Q = charge, t = time

• In a circuit, we can consider that positive charge is repelled by the positive terminal and attracted by the negative terminal, so charge and current flow from positive to negative. This is called the 'conventional current'. In reality, negatively charged electrons flow in the opposite direction, from negative to positive.

Sample question   A charge of 35 C flows round a circuit in 14 s.

a) Work out the current flowing.

b) The charge on each electron is 1.6×10^{-19} C. Work out the number of electrons flowing round the circuit in this time.

[4 marks]

Student's answer   a) $I = \frac{Q}{t} = \frac{35}{14} = 2.5$ A   [2 marks]

b) Number of electrons $= \frac{35}{1.6 \times 10^{-19}} = 2.19 \times 10^{19}$ [1 mark]

Examiner's comments   a) *Correct answer*

b) *The student started the calculation correctly but made a small mistake in the calculation of the powers of 10.*

Correct answer   a) $I = \frac{Q}{t} = \frac{35}{14} = 2.5$ A   [2 marks]

b) Number of electrons $= \frac{35}{1.6 \times 10^{-19}} = 2.19 \times 10^{20}$ [2 marks]

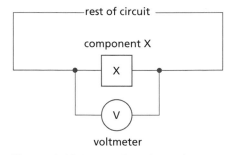

Figure 4.16 Measuring the p.d. across component X

Electromotive force and potential difference

• **Electromotive force** (emf) is the source of electrical energy for a whole circuit. It is measured in volts.

• Emf (in volts) is the amount of energy (in joules) given to each coulomb of charge as it passes round the circuit.

• **Potential difference** (p.d.) across a component is a measure of the electrical energy transferred in the component. It is measured in volts. It is the amount of energy (in joules) given to each coulomb of charge as it passes through the component.

• P.d. is measured by a voltmeter, which must be connected between the two points, **in parallel** with any circuit elements between the points.

• Figure 4.16 shows a voltmeter in parallel with component X.

Common error

✗ Putting a voltmeter **in series** with the part of the circuit where p.d. is to be measured.

✓ The voltmeter must be connected **in parallel**. ∎

● **Try this**   The answers are given on **p. 103**.

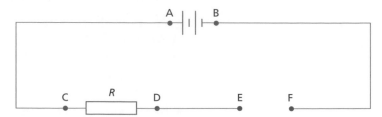

Figure 4.17

5 You are asked to take measurements from the circuit shown in Figure 4.17 and are provided with an ammeter, a voltmeter and a piece of wire. Complete the table below to indicate which component, if any, you should connect across the points AB, CD and EF to take each measurement.

Measurement to be taken	AB	CD	EF
Emf of battery			
Current through *R* when connected to battery			
P.d. across *R* when connected to battery			

Resistance

- For a given p.d., the greater the resistance the smaller the current.
- For a given resistance, the greater the p.d. the greater the current.

> **Examiner's tip**
> ▶ You must be able to recall and use the equation: $R = \dfrac{V}{I}$
>   *R* = resistance in ohms (Ω), *V* = p.d., *I* = current

● **Experiment to determine resistance of unknown conductor Y**

1) Connect the circuit as shown in Figure 4.18.

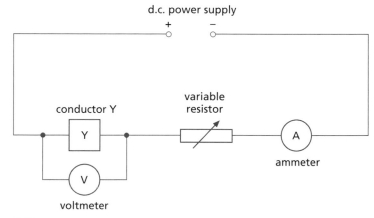

Figure 4.18

2) Ensure the ammeter is **in series** with the power supply and conductor Y (so that the current flows *through* the ammeter).
3) Ensure the voltmeter is **in parallel** with the conductor Y (so that the voltmeter measures the p.d. *between* the ends of the conductor).
4) Record the values of the p.d. in volts and current in amps.
5) Change the setting of the variable resistor and record a series of at least five more pairs of values.
6) Work out the value of the resistance using $R = \dfrac{V}{I}$ for each pair of readings.
7) Work out the average value of *R*, or use a graphical method.

Sample question A student carries out an experiment to find the resistance of a wire using an ammeter and a voltmeter. The table below shows his first three readings.

Reading	1	2	3	4
Voltage/V	1.10	2.10	3.05	3.95
Current/A	0.20	0.35	0.50	?
Resistance/Ω	5.50	6.0	?	–

Make sure you show your working for all parts of this question.
a) Work out the resistance for the third pair of readings. [1 mark]
b) Work out the average value of resistance. [1 mark]
c) Using this average value of resistance, work out the current the student can expect when he takes the fourth reading.
[2 marks]

Student's answer a) Reading 3: $R = \dfrac{V}{I} = \dfrac{3.05}{0.50} = 6.1\,\Omega$ [1 mark]

b) Average $V = \dfrac{(1.10 + 2.10 + 3.05)}{3} = \dfrac{6.25}{3} = 2.08\,V$

Average $I = \dfrac{(0.20 + 0.35 + 0.50)}{3} = \dfrac{1.05}{3} = 0.35\,A$

Average $R = \dfrac{2.08}{0.35} = 5.94\,\Omega$ [0 marks]

c) Reading 4: $I = VR = 3.95 \times 5.94 = 23.47\,A$ [0 marks]

Examiner's comments a) *Correct answer*
b) *The student should have taken the average of the three values of resistance from part a). It is simply good fortune that his answer is close to the correct answer.*
c) *The student has rearranged the formula $R = \dfrac{V}{I}$ incorrectly.*

Observation and comparison with the first three readings show that the calculated value is unlikely.

Correct answer a) Reading 3: $R = \dfrac{V}{I} = \dfrac{3.05}{0.50} = 6.1\,\Omega$ [1 mark]

b) Average $R = \dfrac{(5.50 + 6.00 + 6.10)}{3} = \dfrac{17.6}{3} = 5.87\,\Omega$ [1 mark]

c) Reading 4: $I = \dfrac{V}{R} = \dfrac{3.95}{5.87} = 0.67\,A$ [2 marks]

- The longer the wire, the greater the resistance.
- The thicker the wire, the smaller the resistance.
- Resistance is proportional to length of the wire.
- Resistance is proportional to $\dfrac{1}{\text{cross-sectional area of the wire}}$

Examiner's tips
▶ You must be able to recall and use the above two relationships for resistance.
▶ Many students may find the easiest way to do this is to remember the following formula (even though it is not required by the syllabus):

$$\text{Resistance} = \dfrac{\text{constant} \times \text{length}}{\text{cross-sectional area}}$$

Sample question Sample A is a length of wire of given material.

a) Copy and complete the table below for the resistance of three more samples of wire of the same material. Choose from the words: greater, less, same.

Sample	B	C	D
Length compared with A	×2	same	×2
Diameter compared with A	same	$\frac{1}{2}$	$\frac{1}{2}$
Resistance compared with A			

[3 marks]

b) Add numerical values to your entries in the table to show the order of magnitude of resistance compared with sample A.

[3 marks]

Student's answer

Sample	B	C	D
Length compared with A	×2	same	×2
Diameter compared with A	same	$\frac{1}{2}$	$\frac{1}{2}$
Resistance compared with A	greater ×2	greater ×2	greater ×4

[3 marks]
[2 marks]

Examiner's comment a) *Correct answers – all three samples have greater resistance than sample A.*

b) *The resistance of sample B will be 2 × greater – the student's answer is correct. Resistance varies with the inverse of area not diameter so the answer to C is incorrect. Although the answer to D is incorrect, the student has correctly carried over from the answer to C, so no further marks are lost.*

Correct answer

Sample	B	C	D
Length compared with A	×2	same	×2
Diameter compared with A	same	$\frac{1}{2}$	$\frac{1}{2}$
Resistance compared with A	greater ×2	greater ×4	greater ×8

[3 marks]
[3 marks]

Electrical energy and power

The following expressions are derived from the definition of the volt and the relationship between the amp and the coulomb.

- Volts = joules per coulomb
- Amps = coulombs per second

- Using the symbols E = energy, Q = charge, t = time:

$$V = \frac{E}{Q} \text{ and } I = \frac{Q}{t}$$
$$\rightarrow V = \frac{E}{It}$$
$$\rightarrow E = VIt$$

- Power in watts = joules per second

$$P = \frac{E}{t}$$
$$\rightarrow P = \frac{VIt}{t} = VI$$

Examiner's tip

▶ You must be able to recall and use the relationships:

$E = VIt$ and $P = VI$

Sample question   A travel kettle is designed for international use. With a 230 V supply the power rating is 800 W.

   a) Work out the current with a 230 V supply and the resistance of the element. [2 marks]
   b) Find the current and power output of the kettle when used in North America with a 110 V supply. [3 marks]
   c) Comment on the use of this kettle in North America. [1 mark]

Student's answer   **a)** $I = \frac{P}{V} = \frac{800}{230} = 3.48\,\text{A}$

$R = \frac{V}{I} = \frac{230}{3.48} = 66.1\,\Omega$ [2 marks]

   b) I will stay the same.
   $P = 110 \times 3.48 = 383\,\text{W}$ [1 mark]
   c) The kettle will take longer to boil water. [1 mark]

Examiner's comments   **a)** *Correct answers*
   b) *The current cannot stay the same because R is the same but V is different. As the student calculated the power from the wrong value of I without any further error, one mark was awarded.*
   c) *The student made a valid comment from the calculated value of P.*

Correct answer   **a)** $I = \frac{P}{V} = \frac{800}{230} = 3.48\,\text{A}$

$R = \frac{V}{I} = \frac{230}{3.48} = 66.1\,\Omega$ [2 marks]

   b) The element is the same so R will stay the same.
   $I = \frac{110}{66.1} = 1.66\,\text{A}$
   $P = 110 \times 1.67 = 183\,\text{W}$ [3 marks]
   c) The kettle will take much longer to boil water. [1 mark]

● **Try this**   The answer is given on **p. 103**.

6 A 3 kW electric heater is used to heat up 2.5 kg of water of specific heat capacity 4200 J/kg/°C. The initial water temperature is 16 °C. Work out the temperature after the heater has been switched on for two minutes.

Electric circuits

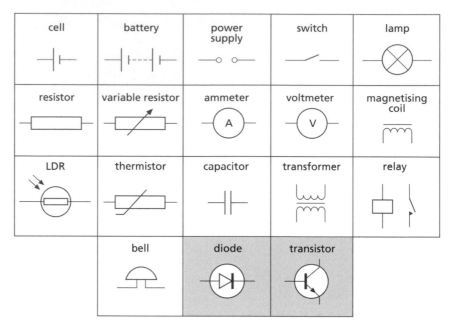

Figure 4.19 Circuit symbols

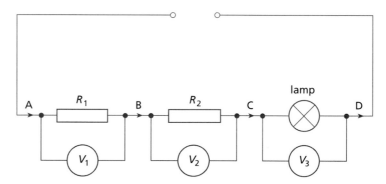

Figure 4.20 Series circuit

In Figure 4.20:

- current at A = current at B = current at C = current at D
- total resistance = R_1 + R_2 + resistance of lamp
- p.d. across the supply = V_1 + V_2 + V_3

Common misconception

✗ Current is 'used up' as it flows round a circuit.
✓ Current is a flow of electrons, which cannot be created or destroyed, so the same number must flow per second through every point in a circuit.
✓ Current is the same at every point in a series circuit. ■

Sample question   In Figure 4.20, $R_1 = 4\,\Omega$ and $R_2 = 3\,\Omega$.
a) Work out the total resistance of R_1 and R_2. [2 marks]
b) The current through R_1 is 1.5 A. State the current through R_2. [2 marks]
c) Work out the voltages V_1 and V_2. [2 marks]
d) The supply voltage is 12 V. Work out the value of V_3. [4 marks]

Student's answer   a) Total resistance $= 4 + 3 = 7\,\Omega$ [2 marks]
b) Current through $R_2 = \frac{3}{4} \times$ current through R_1
$= 0.75 \times 1.5 = 1.125$ A [0 marks]
c) $V_1 = I_1 \times R_1 = 1.5 \times 4 = 6$ V
$V_2 = I_2 \times R_2 = 1.125 \times 3 = 3.375$ V [2 marks]
d) $V_3 = 12 - (4 + 3) = 12 - 7 = 5$ V [1 mark]

Examiner's comments   a) *Correct answer*
b) *The student did not recognise that current stays the same through components in series.*
c) *The student correctly applied his answers from b).*
d) *The student might have been on the right lines and then made errors in substituting resistances instead of voltages. With little working and no explanation it is impossible for the examiner to know.*

Correct answer   a) Total resistance $= 4 + 3 = 7\,\Omega$ [2 marks]
b) Current through $R_2 =$ current through $R_1 = 1.5$ A [2 marks]
c) $V_1 = I \times R_1 = 1.5 \times 4 = 6$ V
$V_2 = I \times R_2 = 1.5 \times 3 = 4.5$ V [2 marks]
d) Supply voltage = total of voltages of rest of circuit = 12 V
$12 = V_1 + V_2 + V_3$
$12 = 6 + 4.5 + V_3$
$V_3 = 12 - (6 + 4.5) = 12 - 10.5 = 1.5$ V [4 marks]

● **Try this**   The answers are given on **p. 103**.

7 In Figure 4.20, $R_1 = 2\,\Omega$, $R_2 = 3\,\Omega$, the resistance of the lamp = 0.5 Ω, and the current through the circuit = 0.8 A. Work out a) the supply voltage, b) the voltage across the lamp.

The lamp is then changed for one with a resistance of 1.5 Ω.
c) Work out the new voltage across the lamp.

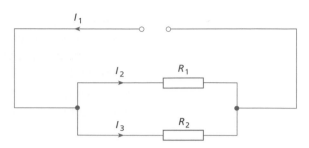

Figure 4.21 Parallel circuit

In Figure 4.21:

- current I_1 from source $> I_2$
- current I_1 from source $> I_3$
- current from source $I_1 = I_2 + I_3$
- combined resistance of R_1 and R_2 in parallel $<$ resistance of R_1
- combined resistance of R_1 and R_2 in parallel $<$ resistance of R_2
- combined resistance of R_1 and R_2 in parallel is given by the formula:

$$\frac{1}{\text{combined resistance}} = \frac{1}{R_1} + \frac{1}{R_2}$$

Sample question

In Figure 4.21, $R_1 = 4\,\Omega$, $R_2 = 3\,\Omega$, $I_1 = 4.2\,\text{A}$ and $I_2 = 1.8\,\text{A}$.
a) Work out the current I_3. [2 marks]
b) Work out the total resistance of R_1 and R_2. [2 marks]
c) Work out the supply voltage. [2 marks]

Student's answer

a) Current $I_3 = I_1 + I_2 = 4.2 + 1.8 = 6.0\,\text{A}$ [0 marks]

b) Total resistance $= \dfrac{1}{R_1} + \dfrac{1}{R_2} = \dfrac{1}{4} + \dfrac{1}{3} = 0.25 + 0.333 = 0.583\,\Omega$ [0 marks]

c) Supply voltage $= 6.0 \times 0.583 = 3.50\,\text{V}$ [2 marks]

Examiner's comments

a) *The student has wrongly thought that I_3 is the total current.*
b) *The student has applied the wrong formula.*
c) *Full marks are given despite the wrong answer. The student correctly followed on from a) and b).*

Correct answer

a) Current $I_3 = I_1 - I_2 = 4.2 - 1.8 = 2.4\,\text{A}$ [2 marks]

b) $\dfrac{1}{\text{total } R} = \dfrac{1}{R_1} + \dfrac{1}{R_2} = \dfrac{1}{4} + \dfrac{1}{3} = 0.25 + 0.333 = 0.583\,\Omega$

Total resistance $= \dfrac{1}{0.583} = 1.72\,\Omega$ [2 marks]

c) Supply voltage $= 2.4 \times 3 = 7.2\,\text{V}$ [2 marks]

● **Try this**   The answers are given on **p. 103**.

8 In Figure 4.21, $R_1 = 2\,\Omega$, $R_2 = 3\,\Omega$, and the supply voltage = 6 V. Work out
   a) I_2, b) I_3, c) I_1, d) the total resistance.

Practical circuits

- In lighting circuits in homes and businesses, lamps are connected in parallel, with these advantages:
   1) Each lamp receives the full mains voltage.
   2) If one lamp should fail, the other lamps will continue to work.

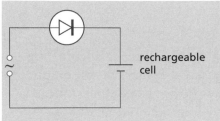

Figure 4.22 Simple charging circuit

- A **diode** only allows current to flow in one direction. It can be used as a rectifier to convert a.c. to d.c. Figure 4.22 shows a diode used as a rectifier to recharge a battery from an a.c. supply.
- A **transistor** can act as a switch linking two circuits, similarly to a relay. A current in one circuit switches on a higher current in another circuit.

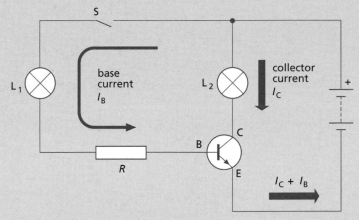

Figure 4.23 Transistor as a switch

Figure 4.23 shows a switching circuit. When S is open, no current flows into the base B of the transistor, so no current flows through the transistor and neither lamp lights. When S is closed, current flows into the base B of the transistor, so current flows through the transistor and lamp L_2 lights. Lamp L_1 does not light because the base current is too small.

- A **potential divider**, or potentiometer, provides a voltage that varies with the values of two resistors in a circuit.

Figure 4.24 shows a potential divider with two separate resistors. If the value of R_1 or R_2 changes, the output voltage will change.

If R_1 increases with R_2 unchanged, the output voltage decreases. If R_2 increases with R_1 unchanged, the output voltage increases.

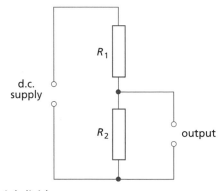

Figure 4.24 Potential divider

• A **rheostat** with a central sliding contact can also be used as a potentiometer.

When the slider moves up and down the resistor in Figure 4.25, the resistance above and below the slider changes so the output voltage changes. As the slider moves down, the output voltage decreases to zero. As the slider moves up, the output voltage increases to the value of the supply voltage.

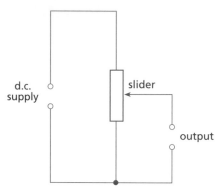

Figure 4.25 Rheostat as a potential divider

• A **thermistor** is a temperature dependent resistor. The resistance falls drastically with increasing temperature. It can be connected in a circuit that is required to respond to change of temperature.

● **Try this**   The answers are given on **p. 103**.

9 In Figure 4.24, the supply voltage is 6 V and R_1 is replaced by a thermistor. The value of R_2 is much less than the greatest resistance of the thermistor and much more than the lowest resistance of the thermistor. What is the output voltage when the thermistor is **a)** very hot, **b)** very cold?

Figure 4.26 shows a circuit that acts as a fire alarm. When the temperature of the thermistor rises, its resistance falls. The thermistor and fixed resistor R are a potential divider, so the voltage of point S rises and enough current flows into the relay for it to switch on the bell.

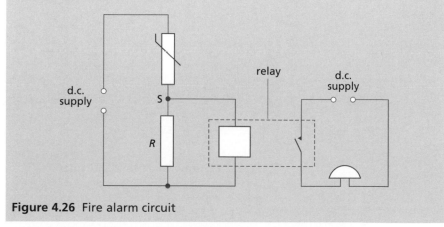

Figure 4.26 Fire alarm circuit

• The resistance of a **light dependent resistor** (**LDR**) falls with increasing light level. It can be connected in a circuit that is required to respond to change of light level.

Figure 4.27 shows a circuit that acts as a warning when too much light enters a photographic lab. With increasing light level, the resistance of the LDR falls. The LDR and fixed resistor *R* are a potential divider, so the voltage of point T rises and enough current flows into the transistor for it to switch on the lamp.

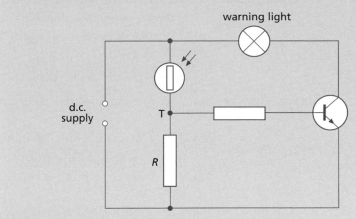

Figure 4.27 Light-sensitive switch using a transistor

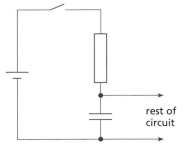

Figure 4.28 Capacitor in time delay circuit

- If a **capacitor** is connected to a d.c. power supply, charge flows into the capacitor and stores energy.

Figure 4.28 shows a capacitor circuit element that provides a time delay. When the switch is closed, it takes some time for the capacitor to charge up to the voltage of the cell. Thus, there is a delay before the full voltage is applied to the rest of the circuit.

Sample question   Figure 4.29 shows a warning circuit.

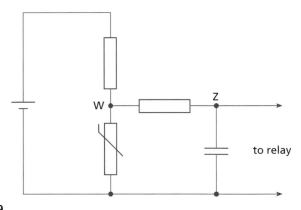

Figure 4.29

The temperature of the thermistor is initially high and then falls quickly by a large amount. The resistance of the fixed resistor is large compared with the initial resistance of the thermistor.

a) State what happens to the resistance of the thermistor.
                                                 [1 mark]/[2 marks]

b) State what happens to the voltage at point W.
                                                 [1 mark]/[2 marks]

c) State what happens to the voltage at point Z.   [2 marks]

d) Describe a suitable use for this circuit.   [2 marks]

Student's answer

a) The resistance of the thermistor increases. [1 mark]/[1 mark]
b) The voltage at point W increases. [1 mark]/[1 mark]
c) The voltage at point Z increases. [1 mark]
d) The circuit could be used as a temperature warning. [0 marks]

Examiner's comments

a) *and* b) *Correct answers for Core students but Extension students need to answer in greater detail.*
c) *The student has not considered the time delay effect of the capacitor.*
d) *The student has given a very vague answer that does little more than re-state the question.*

Correct answer

a) The resistance of the thermistor increases by a large amount.
   [1 mark]/[2 marks]
b) The voltage at point W increases from close to zero to close to the voltage of the cell. [1 mark]/[2 marks]
c) The voltage at point Z increases slowly up to the voltage at point W. [2 marks]
d) The circuit could be used as a warning of a fall in temperature inside a device, e.g. inside an industrial oven or furnace to warn if the heater failed. There is a time delay so there is no warning after temporary decreases, e.g. when opening the door to remove an item. [2 marks]

Digital electronics

- In an **analogue** circuit, the voltages can vary continuously to take any value in the available range.
- In a **digital** circuit, voltage can only be in one of two states:
  1) OFF, LOW or zero voltage (denoted by binary 0)
  2) ON, HIGH or maximum available voltage (denoted by binary 1).
- **Logic gates** are digital components with one or more inputs, which determine the value of the one output. Logic gates contain transistors acting as switches and other components.
- The truth table for a logic gate shows the relationship between the input and output binary values.

Examiner's tip
▶ You must be able to describe the action of the NOT, OR, AND, NOR and NAND gates shown below with their truth tables, as well as state and use their symbols.

- A NOT gate reverses the value of its single input.

Figure 4.30 NOT gate

Input	Output
0	1
1	0

Figure 4.31 OR gate

- An OR gate has a HIGH output if either or both of the inputs are HIGH.

Input A	B	Output
0	0	0
0	1	1
1	0	1
1	1	1

Figure 4.32 AND gate

- An AND gate has a HIGH output only if both of the inputs are HIGH.

Input A	B	Output
0	0	0
0	1	0
1	0	0
1	1	1

Figure 4.33 NOR gate

- A NOR gate is the same logically as an OR gate followed by a NOT gate.

Input A	B	Output
0	0	1
0	1	0
1	0	0
1	1	0

Figure 4.34 NAND gate

- A NAND gate is the same logically as an AND gate followed by a NOT gate.

Input A	B	Output
0	0	1
0	1	1
1	0	1
1	1	0

- Circuits often use a combination of gates. A NOR gate circuit is simple to make with one transistor, so many integrated circuits (microchips) combine large numbers of NOR gates.

Truth table for the circuit in Figure 4.35:

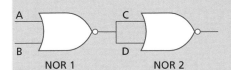

Figure 4.35 Combining two NOR gates to make an OR gate

Inputs to NOR 1 A	B	Output from NOR 1	Inputs to NOR 2 C	D	Output from NOR 2
0	0	1	1	1	0
0	1	0	0	0	1
1	0	0	0	0	1
1	1	0	0	0	1

The whole circuit therefore acts as an OR gate.

Sample question a) Draw the truth table for the circuit shown in Figure 4.36.

[3 marks]

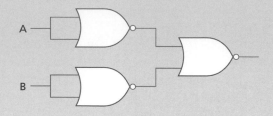

Figure 4.36

b) Which circuit element is equivalent to the circuit shown in Figure 4.36? [1 mark]

Student's answer a)

Input		Output
A	B	
0	0	0
0	1	1
1	0	1
1	1	1

[1 mark]

b) The circuit acts as an OR gate. [1 mark]

Examiner's comments a) *The student failed to label intermediate points in the circuit and include them in the truth table. The second and third lines in the table contain errors and without working the student is given little credit.*

b) *Even though the answer is wrong, the deduction from the student's truth table was correct.*

Correct answer a) Label the inputs to the right-hand NOR gate as C (upper) and D (lower).

Input				Output
A	B	C	D	
0	0	1	1	0
0	1	1	0	0
1	0	0	1	0
1	1	0	0	1

[3 marks]

b) The circuit acts as an AND gate. [1 mark]

● **Try this** The answers are given on **p. 103**.

10 Copy and complete the truth table for the circuit shown in Figure 4.37.

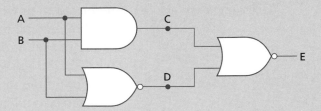

Figure 4.37

A	B	C	D	E
0	0			
0	1			
1	0			
1	1			

Dangers of electricity

- Damaged insulation can lead to danger of electric shock or fire.
- Cables overheated due to excessive current can lead to fire in the appliance or damage to the insulation.
- Water lowers the resistance to earth, so damp conditions can lead to current shorting and can cause shocks. Electrical devices for use in damp conditions must be designed to high standards of damp-proofing, especially connectors and switches.
- A fuse is a piece of wire that melts when too much current flows through it. This switches off the circuit to protect against shock, fire or further damage.
- A circuit breaker is a type of relay which is normally closed, but opens when too much current flows through it. Circuit breakers react much faster than fuses and have increasingly replaced them in consumer units.

Sample question A family is having an evening barbeque on their lawn in a country where dew forms on the grass after dark. Explain whether the following three situations are potentially dangerous.
  a) A heater and several other high-power electrical devices are supplied by an old extension cable. [2 marks]
  b) There is a cut in the outer insulation of the cable. [2 marks]
  c) The devices are connected to a multi-socket lying on the lawn. [2 marks]

Student's answer **a)** The electrical power is likely to require too much current in the cable, leading to overheating. This could cause a fire or melting of the insulation. [2 marks]
  b) There is insulation on the individual wires so the cable is safe. [0 marks]
  c) The dew on the cable connection could cause an electric shock. [2 marks]

Examiner's comments
a) *Correct answer*
b) *Incorrect answer – using a cable with any sort of cut is unsafe practice.*
c) *Correct answer*

Correct answer
a) The electrical power is likely to require too much current in the cable, leading to overheating. This could cause a fire or melting of the insulation. [2 marks]
b) This is dangerous because there could also be a cut in the insulation of the individual wires, which would be difficult to see. There would be a danger of electric shock. [2 marks]
c) The dew on the cable connection could cause an electric shock. [2 marks]

The motor effect

- A wire or conductor carrying a current in a magnetic field experiences a force. This can be demonstrated with the apparatus shown in Figure 4.38.

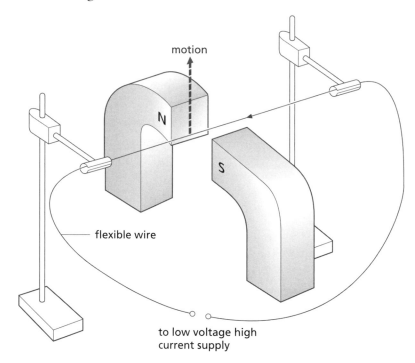

Figure 4.38 Demonstrating the motor effect

The loosely suspended wire will be seen to move up when the current is switched on. The wire will move down if either the current is reversed or the magnet poles are swapped to reverse the field. If both the field and current are reversed, the wire will again move up.

> **Examiner's tips**
> ▶ You must be able to state and use **Fleming's left-hand rule** to determine the relative directions of field, current and force. Hold your thumb and first two fingers of your left hand at right angles to one another. The thu**M**b gives the direction of the force (or **M**otion), the **F**irst finger the **F**ield and se**C**ond finger the **C**urrent.
> ▶ Use the letters in bold capital letters to remember which is which.

Figure 4.39 shows a coil in a magnetic field carrying a current.

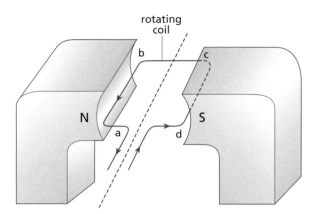

Figure 4.39 The principle of a simple d.c. motor

There is a force from the magnetic field on the wire **ab** in one direction (upwards). The current in the wire **cd** is in the other direction so it experiences a force in the opposite direction (downwards). These two forces cause the coil of wire to turn. The turning effect is increased by:

1) increasing the number of turns on the coil
2) increasing the current.

This turning effect causes the coil of a d.c. motor to rotate continuously. There is a switching mechanism (commutator and brushes) that changes the direction of the current every half turn to allow continuous rotation.

Electromagnetic induction

● When the magnetic field through a circuit changes, an emf (voltage) is induced.

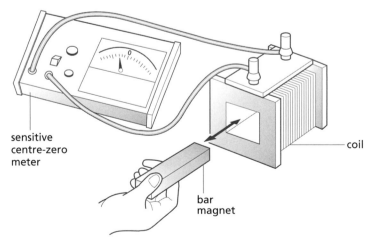

Figure 4.40 A current is induced in the coil when the magnet is moved in or out

● The induced emf increases with an increase of:
  1) speed of relative motion of the magnet and circuit
  2) number of turns on the coil
  3) strength of the magnet.

• The direction of the induced emf *opposes* the change which caused it. In Figure 4.41, the moving magnet induces an emf in the coil, which causes a current to flow through the coil. The current produces a magnetic field, which opposes the movement of the magnet.

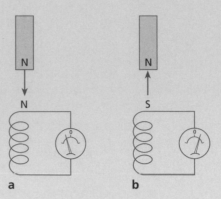

Figure 4.41 The induced current opposes the motion of the magnet

• Figure 4.42 shows a simple **a.c. generator** with a coil rotating within the field of a magnet. An emf is induced in the coil connected to the output terminals through slip rings on the axle, which rotate with the coil, and fixed carbon brushes. The wires on each side of the coil cut the field alternately moving up and down, so the emf is induced in alternating directions, as shown by the graph in Figure 4.43.

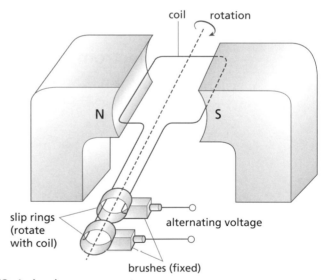

Figure 4.42 A simple a.c. generator

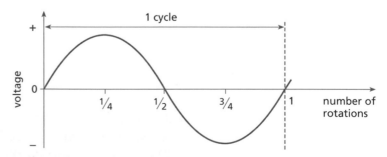

Figure 4.43 Output from a simple a.c. generator

- Figure 4.44 shows a **transformer** with two coils wound on the same soft iron core. The primary coil is supplied with an a.c. and the secondary coil provides a.c. to another circuit. The primary coil acts as an electromagnet and produces an alternating magnetic field in the soft iron core. The secondary coil is in this alternating magnetic field, so an alternating emf is induced.

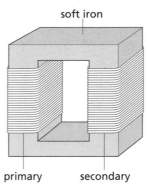

soft iron

primary   secondary

Figure 4.44 Primary and secondary coils of a transformer

- Transformers only work with a.c. and change the voltage from one value to another.

Common misconception

✗ Transformers work with a.c. and d.c.
✓ Transformers only work with a.c. ■

Examiner's tips

▶ You must be able to recall and use the equation $\frac{V_p}{V_s} = \frac{N_p}{N_s}$ where V_p and V_s are the voltages across the primary and secondary coils and N_p and N_s are the numbers of turns of the primary and secondary coils.

▶ You must be able to recall and use the equation $V_p I_p = V_s I_s$ for a 100% efficient transformer, where I_p and I_s are the currents flowing in the primary and secondary coils.

- Electricity is transmitted over large distances at very high voltages in order to reduce the energy losses due to the resistance of the transmission lines. This is achieved by having a step-up transformer at the power station to increase the voltage to several hundred thousand volts. Where the electricity is to be used, there is a series of step-down transformers to reduce the voltage to values suitable for use in factories or homes.

- Power loss in a cable = p.d. across length of cable × current (I)

For a given cable of fixed resistance,

p.d. = I × resistance of cable (R)

Power loss = $I \times R \times I = I^2 R$

Power is therefore transmitted at the highest possible voltage in order to reduce the current and thus the losses in the cables.

• **Try this** The answer is given on **p. 103**.

11 A transformer is used to provide an a.c. 6 V supply for a laboratory from 240 V a.c. mains. The secondary coil of the transformer has 100 turns. Work out how many turns the primary coil should have.

Cathode rays

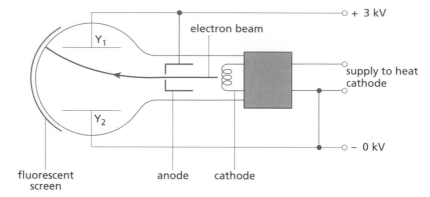

Figure 4.45 Production and deflection of cathode rays

- Figure 4.45 shows a tube used to produce cathode rays. There is a vacuum in the tube. A **cathode ray** is a beam of electrons given off by a hot cathode (called 'thermionic emission' as it is caused by heat). Electrons are attracted to the positive anode; some pass through the hole in the anode and travel the length of the tube. A fluorescent screen glows when struck by electrons, showing their position.
- With no voltage between the plates Y_1 and Y_2 the electrons pass straight on without deflection. If a large voltage is applied between the plates, with Y_1 positive, there is an electric field between them, the negative electrons are attracted to Y_1 and the beam is deflected upwards as shown.

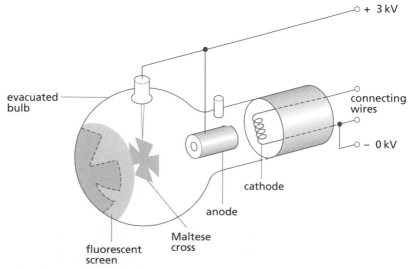

Figure 4.46 Maltese cross tube

• A beam of charged particles is equivalent to a current and experiences a similar force in a magnetic field. Figure 4.46 shows an experiment with a beam of electrons in a Maltese cross tube. Those that miss the cross cause the fluorescent screen to glow, showing a shadow pattern in the shape of the cross. If two bar magnets are placed each side of the tube to produce a horizontal magnetic field, the shadow pattern moves up or down as the electrons in the beam experience a force from the magnetic field.

Common misconception

✗ A beam of moving electrons is equivalent to a current flowing in the same direction as the electrons.

✓ Electrons are negatively charged so the current is in the *opposite* direction to the electron flow.

✓ A beam of positively-charged particles is equivalent to a current flowing in the same direction as the motion of the particles. ■

Sample question   Figure 4.47 shows the fluorescent screen of an electron tube and the shadow of a Maltese cross when electrons travel without any magnetic field.

Figure 4.47

State and explain how this shadow will move when two bar magnets are placed in the positions shown.   [4 marks]

Student's answer   The magnetic field is from left to right. [1 mark] The electrons move out of the paper. [0 marks] Using the left-hand rule, the beam and shadow move upwards. [2 marks]

Examiner's comments   *The student's statement about the magnetic field is correct but electrons are negatively charged so when they move in one direction, the corresponding current is in the opposite direction. The student correctly applied the left-hand rule to their directions, even though the current direction was wrong.*

Correct answer   The magnetic field is from left to right. [1 mark] The electrons move towards the observer so the current moves into the paper. [1 mark] Using the left-hand rule, the beam and shadow move downwards. [2 marks]

Cathode ray oscilloscope

- The cathode ray oscilloscope (CRO) is an important scientific instrument used to display a voltage that varies with time, especially waveforms whose oscillations can be converted into a variable voltage by a sensor.

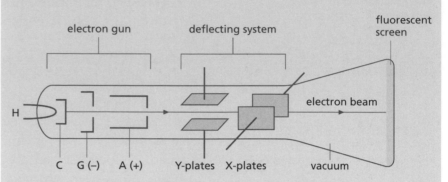

Figure 4.48 Main parts of a CRO

- The operation of a CRO can be seen in Figure 4.48. The heater H causes thermionic emission of electrons from the cathode, C. The anode A is at a high positive voltage relative to C and accelerates the electrons along the tube. The grid G is at a negative voltage relative to C and acts as a brightness control by controlling the number of electrons passing through. The X and Y plates deflect the electron beam horizontally and vertically to show the waveform, which is displayed when the electrons cause the fluorescent screen to glow.
- An internal circuit provides a voltage increasing with time (called the time base), which is applied to the X plates so the beam moves steadily across the screen.
- The voltage from the sensor is applied to the Y plates so the beam moves up and down to show the displacement of the wave.

TOPIC 5 Atomic Physics

Key objectives

- To know and be able to describe the structure of an atom and its nucleus
- To be able to use the term 'isotope'
- To be able to use nuclide notation and write radioactive decay equations using this notation
- To know the basic characteristics of the three types of radioactive emissions
- To be able to perform simple half-life calculations
- To be able to describe precautions for handling, using and storing radioactive materials safely

Key definitions

Term	Definition
Atom	The smallest possible unit of an element
Isotope	One form of an element that has a different number of neutrons in the nucleus from other isotopes of the same element
α-particle	A helium nucleus made up of two protons and two neutrons
β-particle	A high-speed electron emitted by a nucleus
γ-ray	A high-frequency electromagnetic wave
Half-life	The average time for half the atoms in a radioactive sample to decay

Key ideas

Atomic model

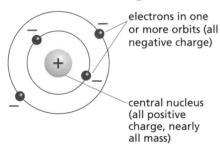

electrons in one or more orbits (all negative charge)

central nucleus (all positive charge, nearly all mass)

Figure 5.1 The nuclear atom

- The atom is the smallest particle of an element. It is made up of a central nucleus, with all the positive charge and nearly all the mass, and electrons in orbit. The nucleus is very much smaller than the electron orbits so the majority of every atom is empty space.
- The nuclear atom was confirmed by observing that although the vast majority of a beam of α-particles (positively-charged subatomic particles) travelling towards a thin metal foil passed straight through without being deflected, a few α-particles were deflected, some through a large angle, and a very small proportion bounced back (see Figure 5.2).

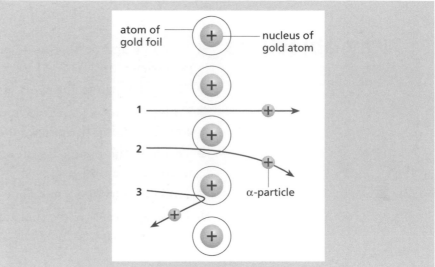

Figure 5.2 Scattering of α-particles by thin gold foil

Path 1 is a long way from any nucleus and is undeflected.
Path 2 is close to a nucleus and there is some deflection.
Path 3 heads almost straight for a nucleus and rebounds back.

Nucleus

- A nucleus contains protons and neutrons, together called **nucleons**.
- A proton is about 2000 times more massive than an electron and has a positive charge of the same size as an electron's negative charge.
- A neutron has about the same mass as a proton but no charge.

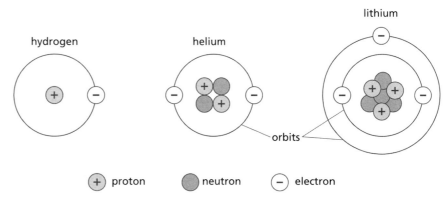

Figure 5.3 Protons, neutrons and electrons in atoms

- The number of protons in a nucleus is called the **proton number** (Z) and is the same as the number of electrons in orbit.
- The number of neutrons is about the same as the number of protons for light elements and rises to about one and a half times the number of protons for the heaviest elements.
- The number of nucleons (protons and neutrons) is called the **nucleon number** (A).
- The nuclide (type of nucleus) of an element can be written with the notation $^{A}_{Z}X$, where X is the chemical symbol for the element.

Isotopes

- Isotopes of the same element are different forms which have the same number of protons in the nucleus but different numbers of neutrons.

Sample question

The normally occurring isotope (version) of carbon is carbon-12, written $^{12}_{6}C$ in nuclide notation.

a) Write down the nucleon and proton numbers of carbon-12.
[2 marks]

b) Write down the number of electrons in a carbon-12 atom.
[1 mark]

c) Carbon-14 is a radioactive isotope (version) that exists in small quantities in the atmosphere. Write down the nucleon and proton numbers of carbon-14. [1 mark]

d) Write down the nuclide notation for carbon-14. [2 marks]

e) Work out the number of neutrons in a nucleus of carbon-14 and state the difference between the nuclei of carbon-12 and carbon-14. [2 marks]

Student's answer

a) Nucleon number, A = 12
Proton number, Z = 6 [2 marks]

b) Number of electrons = 12 [0 marks]

c) Nucleon number, A = 14
Proton number, Z = 6 [1 mark]

d) $^{6}_{14}C$ [0 marks]

e) A carbon-14 nucleus is bigger, with more particles. [0 marks]

Examiner's comments

a) *Correct answers*

b) *The student has incorrectly thought that the number of electrons is the same as the number of nucleons.*

c) *Correct answers*

d) *The student has mixed up the nucleon and proton numbers – care needs to be taken here!*

e) *The first part of the student's answer has been omitted and the rest of the answer is much too vague.*

Correct answer

a) Nucleon number, A = 12
Proton number, Z = 6 [2 marks]

b) Number of electrons = number of protons = 6 [1 mark]

c) Nucleon number, A = 14
Proton number, Z = 6 [1 mark]

d) $^{14}_{6}C$ [2 marks]

e) Number of neutrons in carbon-14 nucleus
= nucleon number – proton number = $A - Z$ = 14 – 6 = 8

Number of neutrons in carbon-12 nucleus = 12 – 6 = 6

A carbon-14 nucleus has two extra neutrons. [2 marks]

Common errors

✘ Mixing up the positions of the nucleon and proton numbers in nuclide notation.

✔ $_2^4$He has nucleon number, $A = 4$ and proton number, $Z = 2$.

✘ Writing one or both of the numbers to the right of the element symbol, as in some old-fashioned books. (Although you would not be penalised for doing this in an exam.)

✔ $_2^4$He has both numbers to the left of He. ∎

- Some radioactive isotopes occur naturally, e.g. carbon-14 is produced in the atmosphere by cosmic rays.
- Many radioactive isotopes are produced artificially in nuclear reactors and have a wide range of practical uses, e.g. as a source of radiation to kill cancers or as tracers in the human body or in a pipeline.

Examiner's tip
▶ You must be able to state and explain examples of practical applications of radioactive isotopes.

● **Try this** The answers are given on **p. 103**.

1 Copy and complete the table to indicate the composition of an atom of each of the isotopes (versions) of strontium given.

Isotope	Number of protons	Number of neutrons	Number of electrons
$_{38}^{88}$Sr			
$_{38}^{90}$Sr			

Radioactivity

- Radioactivity occurs when an unstable nucleus decays and emits one or more of the three types of radiation: α-(alpha) particles, β-(beta) particles or γ-(gamma) rays.
- In collisions between radioactive particles and molecules in air, the radioactive particles knock electrons out of the atoms, leaving the molecules positively charged. This is called **ionisation**.
- Background radiation occurs naturally, mainly due to radioactivity in rocks and the air and due to particles from space called cosmic rays.
- Radioactivity is a random process. It is impossible to know when an individual radioactive nucleus will decay. A large sample will, on average, decay in a regular pattern but there will always be short-term variations in the rate of decay.
- Radioactivity can be detected by a Geiger–Müller (GM) tube, as shown in Figure 5.4. Radiation causes a pulse of current to flow between the electrodes, which flows to the counter or ratemeter and often to a loudspeaker to produce the characteristic 'clicking'.

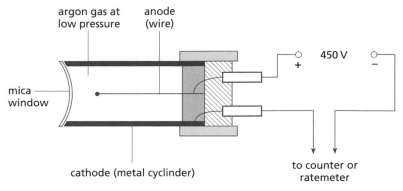

argon gas at low pressure

anode (wire)

450 V

mica window

cathode (metal cyclinder)

to counter or ratemeter

Figure 5.4 GM tube

α-particles, β-particles and γ-rays

Emission	Nature	Charge	Penetration	Ionising effect
α-particle	Helium nucleus (two protons and two neutrons)	+2	Stopped by thick paper	Very strong
β-particle	High-speed electron	−1	Stopped by a few millimetres of aluminium	Weak
γ-ray	Electromagnetic radiation	none	Only stopped by many centimetres of lead	Very weak

● **Try this** The answers are given on **p. 103**.

2 The penetration of two radioactive samples is tested by measuring the count rate with various types of shielding between the sample and the counter. The numbers in the table below indicate the count rate (CR) with each type of shielding in place – no shielding, thick card, 3 mm of aluminium (Al), 3 cm of lead (Pb). Copy the table and tick the appropriate boxes in the right-hand three columns to show the type or types of emission from that sample.

Sample	CR (none)	CR (card)	CR (Al)	CR (Pb)	α	β	γ
1	6000	1000	1000	20			
2	3000	3000	20	20			

- The very strong ionising effect of α-particles causes their short range. They only travel a few centimetres in air, as they collide with air molecules and ionise them.
- β-particles travel a few metres in air, as they ionise the air molecules much less.
- γ-rays do not directly ionise the air but can cause emission of electrons, which behave as β-particles.

Deflection of radioactive emissions in magnetic and electric fields

- α-particles are positively charged and are deflected, as an electric current flowing in the same direction.
- β-particles are electrons and are deflected in the same way as cathode rays.

- γ-rays are electromagnetic waves and are not deflected (see Figure 5.5).

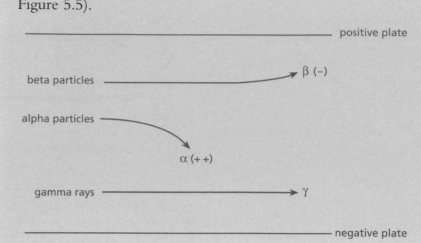

Figure 5.5 Deflection of α-and β-particles and γ-rays in an electric field

Try this The answers are given on **p. 103**.

3 A positron is a sub-nuclear particle with the same mass as an electron but a positive charge. A certain nuclear reaction emits positrons and γ-rays, which are both directed to pass parallel and between two horizontal plates in a vacuum. The upper plate has a very high positive potential relative to the lower plate. Describe the path between the plates of **a)** the positrons **b)** the γ-rays.

Radioactive decay

- In α-decay, the nucleus loses two neutrons and two protons. The nucleon number goes down by four. The proton number goes down by two, so the nuclide changes to another element. Two orbital electrons drift away to make the atom electrically stable and the new number of electrons equals the new proton number. An example of α-decay can be shown by a word equation:

radium-226 → radon-222 + α-particle

The same example of α-decay can be shown by an equation in nuclide notation:

$$^{226}_{88}\text{Ra} \rightarrow {}^{222}_{86}\text{Rn} + {}^{4}_{2}\text{He}$$

Note that because an α-particle is the same as a helium nucleus, it is shown as 'He' in nuclide notation.

- In β-decay, a neutron in the nucleus changes to a proton and an electron, which is emitted at high speed as a β-particle. The nucleon number is unchanged. The proton number goes up by one, so the nuclide again changes to another element. An electron is collected from the surroundings to make the atom electrically neutral and the new number of orbital electrons equals the new proton number. An example of β-decay can be shown by word and nuclide equations:

carbon-14 → nitrogen-14 + β-particle

$$^{14}_{6}\text{C} \rightarrow {}^{14}_{7}\text{N} + {}^{0}_{-1}\text{e}$$

Note that because a β-particle is an electron, it is shown as 'e' in nuclide notation with a nucleon number of 0, because it has negligible mass, and a proton number of –1, because of its negative charge.

- γ-rays are usually given off during both α-decay and β-decay, and can be added to the equations, for example:

$$^{226}_{88}\text{Ra} \rightarrow {}^{222}_{86}\text{Rn} + {}^{4}_{2}\text{He} + \gamma$$

Common misconception

✗ α-particles and β-particles can be given off in the same decay reaction.
✓ α-particles and γ-rays are possible together.
✓ β-particles and γ-rays are possible together. ■

Sample question Radioactive strontium-90 (Sr, proton number 38) decays to yttrium (Y), emitting a beta particle and gamma rays. Show this decay reaction as a nuclide equation. [4 marks]

Student's answer $^{90}_{38}\text{Sr} \rightarrow {}^{89}_{39}\text{Y} + {}^{0}_{-1}\text{e} + \gamma$ [3 marks]

Examiner's comments *The student has mostly got the answer right. The nuclide symbol for strontium is entirely correct, as are the symbols for the beta particle and gamma rays. The student has also correctly deduced that the proton number of yttrium is 39, one more than that of strontium. However, there is confusion in finding the nucleon number, which must stay the same in beta decay.*

Correct answer $^{90}_{38}\text{Sr} \rightarrow {}^{90}_{39}\text{Y} + {}^{0}_{-1}\text{e} + \gamma$ [4 marks]

● **Try this** The answer is given on **p. 103**.

4 Radioactive uranium-238 (U, proton number 92) decays to thorium (Th), emitting an alpha particle and gamma rays. Show this decay reaction as a nuclide equation.

Half-life

- Radioactive decay is a random process. It is impossible to predict when an individual nucleus will decay. However, *on average*, there is a definite decay rate for each isotope.
- The decay rate is expressed as the half-life, which is the time for half the atoms in a sample to decay. As the process is random, there may be some small fluctuations but with a large number of atoms in even a small sample, the half-life is effectively constant.

Examiner's tips
▶ Strong electric, magnetic or gravitational fields, changes in temperature, aggressive chemical reagents, or other elements that a radioactive isotope is combined with chemically, all have no effect on the value of its half-life.
▶ You must be able to understand and use the term 'half-life' in simple calculations with data in graphs or tables.

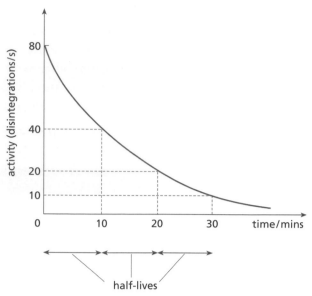

Figure 5.6 The half-life of a material can be found by using a graph (decay curve)

Sample question A radioactive sample gives a Geiger–Müller tube reading of 700 counts per second, including a background count of 100 counts per second. The half-life of the sample is seven days.

a) Work out the expected Geiger–Müller tube reading three weeks later. [4 marks]

b) The value actually recorded three weeks later was 8 counts per second different from the expected value. Explain why this might be so. [2 marks]

Student's answer a) After one week, GM tube reading = $\frac{700}{2}$ = 350 counts/s

After two weeks, GM tube reading = $\frac{350}{2}$ = 175 counts/s

After three weeks, GM tube reading = $\frac{175}{2}$ = 87 counts/s [2 marks]

b) The difference could have been caused by the sample warming up. [0 marks]

Examiner's comments a) *The student has failed to consider the background radiation, which does not decay.*

b) *The temperature of a sample has no effect on its half-life.*

Correct answer a) Initial GM tube reading due to sample
= 700 − 100 = 600 counts/s

After one week, GM tube reading due to sample
= $\frac{600}{2}$ = 300 counts/s

After two weeks, GM tube reading due to sample
= $\frac{300}{2}$ = 150 counts/s

After three weeks, GM tube reading due to sample
= $\frac{150}{2}$ = 75 counts/s

Final GM tube reading, including background
= 75 + 100 = 175 counts/s [4 marks]

b) The difference could have been caused by a variation in the background radiation. [2 marks]

✗ Stating that a radioactive sample loses the same number of atoms in the second half-life as in the first.

✓ If a sample of 10 000 atoms has a half-life of one hour, 5000 atoms decay in the first hour and 2500 decay in the second hour. ■

Safety precautions

- Whenever possible, radioactive samples should be in sealed casings so that no radioactive material can escape.
- Samples should be stored in lead-lined containers in locked storerooms.
- Samples should only be handled by trained personnel and must always be supervised when not in store.
- Radioactive samples should be kept at as great a distance as possible from people. In the laboratory, they should be handled with long tongs and students should keep at a distance. In industry, they are usually handled by remote-controlled machines.
- Workers in industry are often protected by lead and concrete walls, and wear film badges that record the amount of radiation received.
- Research, educational and industrial establishments using radioactive sources should have appropriate security procedures in place to prevent abuse by mischievous, criminal or terrorist intruders.

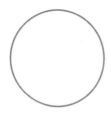

Preparing for the examination

During the course

Preparing for an external examination is a continuous process throughout the course. All the activities, lessons, homework, practical work and assessments are major factors in determining your final examination grade, so the first piece of advice is to suggest that you work steadily throughout the two years of your course. It is essential that you prepare thoroughly for internal school examinations, then, as you approach the IGCSE examination and start your revision programme, the topics will be familiar and the learning process will be less stressful and more productive.

Make sure that your notes are up-to-date. If you miss work through absence, either copy it from a reliable friend or leave a comment in your notes that will remind you to refer to the topic in a textbook. Similarly, look at any homework you have missed and if it involves the reinforcement of skills or concepts, then it would be a good idea to photocopy it. This may not be a popular suggestion but you must realise that not all homework assignments are of equal value – some are integral components of the course. You will need to make a judgement, but to help you here is some general advice: problems, problem solving, factual material that is not covered elsewhere in your scheme of work and past questions should be photocopied or, even better, you should complete the assignments on your return.

In summary:

- work steadily throughout the course
- ensure that your work is both complete and accurate
- learn the topics for tests and internal examinations
- if you find an aspect of the course difficult seek assistance.

Revision techniques

Well in advance of the examination, produce a revision timetable for all your subjects. Be realistic – you must include time for relaxation and socialising. Then create a more detailed one for Physics, to cover all the topics. Ideally, you ought to go through the complete course twice. Keep a checklist of the topics studied – it is encouraging to have a visual record of your progress. It is useful to have a copy of the syllabus but not essential, as this book includes all the information required for IGCSE Physics.

Ideally, you should have a quiet room at a comfortable temperature, plenty of paper and a pencil or pen. Some students find doodling (Physics of course) with coloured pens helpful. On occasions, revising with a friend makes a welcome and useful change. You will have to discover for how long you can profitably study. This is a very individual characteristic and can vary from person to person, and may be as little as 30 minutes. Few people can revise solidly and effectively for much more than an hour at a time. Do not exceed your revision time, and break up the available time into study sessions and breaks. You may want to introduce rewards – 'When I have finished this section of work I will …'.

Revision, like all learning, must be active. Do not believe that just looking at a book is an effective way of learning. You could make flash cards that have bullet lists of essential points. You could study the topic for several minutes and then close the book and write out what you can remember, then check your account against the book. Repeat this until you have most of the information correct, then move on to another section of work. This is the 'look, cover, write and check technique' and it is very effective for the majority of students. It is crucial that you repeat this technique on the same topic, at least once but preferably twice, soon afterwards, that is, either later the same day or the next day. This will greatly increase your long-term retention of the topic.

Keeping a list of important words is useful. You could write the words on one side of a card and the meanings on the other. Then when you have an odd few minutes you can brush up on your Physics vocabulary.

Once you have acquired a reasonably good knowledge and understanding of the course, it is time to extend your revision to practising on past papers. This is a most valuable form of preparation; not only does it provide a test of the effectiveness of your revision but it also provides an insight into what to expect in the 'real' examination.

The examination syllabus gives a full list of the terms used by examiners and how candidates are expected to respond. The most common terms are listed on p. viii of this book.

How to approach the examination

If the examination centre has provided a detailed timetable, then highlight your examinations and put the timetable in a prominent place in your home. Ask one of your family to check with you each day your commitments for the next day. This will avoid the following type of situation occurring:

Student 'I thought that the examination was this afternoon.'

Teacher 'No, it was this morning. You will now have to sit it in November or next year.'

Put out the correct equipment the night before – pencil, sharpener, eraser, ruler, protractor, pair of compasses, calculator (are the batteries OK?) and two pens.

Leave home in generous time; if you are late you will not be given extra time and, under certain circumstances, you will not be allowed to enter the examination room. The regulations vary depending on the Examination Board. Do not put yourself at a disadvantage. It is no good saying, for example, 'The 8.45 bus didn't run this morning and I had to wait for the 9.15.' It is the candidate's responsibility to arrive on time for an examination. You should *always* allow time to spare.

Multiple choice papers

Attempt all of the questions as there are no penalties for incorrect responses. If a calculation is involved, be careful to work logically. Beware – one of the incorrect answers may be what you would get with a 'typical' mistake in working out. Never make a blind guess, but try to eliminate some of the incorrect answers to increase the odds in your favour.

Read the question carefully, remembering that at least one of the incorrect answers, called 'distractors', will seem to be correct. Do not rush – think.

If you cannot answer a question, or remain uncertain as to which is the correct answer, leave it and return to it when you have completed the other questions. Use any spare time at the end of the examination to check your answers.

Theory papers

Once the examination has started, flick through the paper and choose a question you feel confident about. You do not have to start with question 1. Read the question twice, look at the mark allocation for each part and then decide exactly what is required to be awarded the marks. This needs a disciplined approach; far too many candidates write at length without answering the question. Never forget – marks are not awarded for correct Physics but for **correct Physics that answers the question**.

- Follow the instructions in the question, being particularly careful to respond to words and phrases such as 'describe' and 'give a reason for'.
- Take reasonable care that your writing is legible – what cannot be read cannot be marked.
- Do not offer more alternative answers than are required in the question, in the vain hope that the examiner will pick out the correct ones. In fact, if only one answer is required and you give two answers, one right and one wrong, the examiner will *always* mark it wrong.
- Avoid 'waffle', as this wastes your time. Similarly, do not re-write the information given in the question and expect to gain marks.
- Do not rush. This is a major cause of mistakes, particularly of misreading the question. The time allocated to the examination is adequate for candidates to complete the paper.
- Leave the hardest questions or parts of questions until last.
- If you finish early, take the opportunity to check through your answers. Ask yourself, 'Have I answered the question and have I made sufficient points to be awarded the marks?'

After the examination, the papers are sent to the examiner allocated to your centre. This examiner will be part of a team headed by a Chief Examiner. All the members of the examining team will look at a sample of their scripts and assess the range of candidates' responses to each question. Then the team will meet to coordinate the marking. For each question, they will decide the range of responses that are acceptable. During the marking period, the Chief Examiner will sample the marking of each examiner, at least twice, to ensure comparability of marking across the team. The scripts and the marks are returned to the Examination Board where the minimum mark for each grade is decided. A few weeks later you are informed of your grade.

Practical examinations

There are three ways of assessing practical skills.

● **School-based assessment**   A series of assessments are conducted throughout the course and should be an integral part of the teaching programme. The teachers will have received guidance and training about the conduct and content of these assessments. The following skills will be assessed:

C1 Using and organising techniques, apparatus and materials
C2 Observing, measuring and recording
C3 Handling experimental observations and data
C4 Planning, carrying out and evaluating investigations

Near the end of course, the results of these assessments are sent to the Examination Board where they are moderated to the same standard across all the centres.

● **Practical test**   This is a single practical test set by the Examination Board and conducted at your centre. The procedures assessed are the same as those in the school-based assessment. This paper is marked by external examiners.

● **Written test of practical skills**   This is a part of the final examination programme. It is a single written paper with a complete emphasis on laboratory procedures. The syllabus gives a full list of the required procedures, a selection of which is:

- recording readings from diagrams of apparatus
- completing tables of data
- drawing conclusions
- plotting graphs
- identifying sources of error and suggesting improvements in experimental procedures.

These papers are marked by a team of external examiners in the same way as the theory papers.

Preparation for practical assessment

The best preparation for the written test is to study some of the past papers, to become familiar with the type of question set. It is likely that the questions on your paper will be similar. Although it is a written test, the practical lessons at school will have provided both the skills and the knowledge needed for this examination.

The two direct assessments of practical skills, that is, the school-based assessment and the practical test, present a different problem from a purely written assessment – in a word, nerves. In any examination you need to be calm and measured in your approach. This is easier to achieve in the written examination – a couple of deep breaths, pick a question you are confident about and 'off you go'. If you make a mistake, you can just delete it and write the correct version (with a reference to the examiner) in a convenient place. This is not the case in practical work. If you make a mistake, you will probably have to start the exercise again. You will have wasted time and not improved your state of mind. When carpenters are being trained to cut pieces of wood to length they are told to 'measure twice then cut once'. This translated into the context of a practical examination is 'Think about what you are about to do. When you are certain of the correct action then carry it out. Do not rush.'

How to improve your grade

Here are a few brief summary points, all of which have been mentioned elsewhere in this book.

- Use this book – it was written to help students attain high grades.
- Learn all the work. Low grades are nearly always attributable to inadequate preparation. If you can recall the work you will succeed; if you cannot you will fail. Harsh but true.
- Practise skills – calculations, problem solving and interpretation of graphs.
- Use past papers to reinforce revision, to become familiar with the types of questions and to gain confidence.
- Answer a question as written on the paper. Do not accept a question as an invitation to write about the topic.

Useful equations

**Topic 1
General Physics**

$$\text{speed} = \frac{d}{t} \qquad \text{speed} = \frac{\text{distance}}{\text{time}}$$

$$\rho = \frac{m}{V} \qquad \text{density} = \frac{\text{mass}}{\text{volume}}$$

$$a = \frac{\Delta v}{t} \qquad \text{acceleration} = \frac{\text{change of velocity}}{\text{time}}$$

$$F = ma \qquad \text{force} = \text{mass} \times \text{acceleration}$$

$$\text{moment} = Fd \qquad \text{moment} = \text{force} \times \text{perpendicular distance from pivot}$$

$$\Delta W = Fd \qquad \text{work done} = \text{force} \times \text{distance moved in direction of force}$$

$$\Delta W = \Delta E \qquad \text{work done} = \text{energy transferred}$$

$$P = \frac{E}{t} \qquad \text{power} = \frac{\text{work done (or energy transferred)}}{\text{time}}$$

$$p = \frac{F}{A} \qquad \text{pressure} = \frac{\text{force}}{\text{area}}$$

$$p = \rho gh \qquad \text{pressure} = \text{density} \times \text{acceleration due to gravity} \times \text{depth below surface of liquid}$$

$$\text{k.e.} = \frac{1}{2} mv^2 \qquad \text{kinetic energy} = \frac{1}{2} \times \text{mass} \times \text{velocity}^2$$

$$\text{p.e.} = mgh \qquad \text{potential energy} = \text{mass} \times \text{acceleration due to gravity} \times \text{change in height}$$

**Topic 2
Thermal Physics**

$$pV = \text{constant} \qquad \text{pressure} \times \text{volume} = \text{constant (at } constant \text{ temperature)}$$

$$l = \frac{\Delta E}{m} \qquad \text{specific latent heat} = \frac{\text{energy supplied}}{\text{mass changing state}}$$

$$c = \frac{\Delta E}{m \, \Delta T} \qquad \text{specific heat capacity} = \frac{\text{energy supplied}}{\text{mass} \times \text{temperature rise}}$$

$$HC = \frac{\Delta E}{\Delta T} \qquad \text{heat (thermal) capacity} = \frac{\text{energy supplied}}{\text{temperature rise}}$$

Topic 3 Waves

$$v = f\lambda \qquad \text{velocity} = \text{frequency} \times \text{wavelength}$$

$$n = \frac{v_{\text{air}}}{v_{\text{glass}}} \qquad \text{refractive index} = \frac{\text{speed of light in air}}{\text{speed of light in glass}}$$

$$\frac{\sin i}{\sin r} = n \qquad \frac{\text{sine angle of incidence}}{\text{sine angle of refraction}} = \text{refractive index}$$

**Topic 4 Electricity
and Magnetism**

$$R = \frac{V}{I} \qquad \text{resistance} = \frac{\text{voltage}}{\text{current}}$$

$$R_s = R_1 + R_2 \qquad \text{series resistance} = \text{resistance 1} + \text{resistance 2}$$

$$\frac{V_p}{V_s} = \frac{N_p}{N_s} \qquad \frac{\text{primary voltage}}{\text{secondary voltage}} = \frac{\text{primary turns}}{\text{secondary turns}}$$

$$I = \frac{Q}{t} \qquad \text{current} = \frac{\text{charge}}{\text{time}}$$

$$R \propto l \qquad \text{resistance is proportional to length of wire}$$

$$R \propto \frac{l}{A} \qquad \text{resistance is proportional to} \frac{1}{\text{cross-sectional area}}$$

$$E = VIt \qquad \text{energy} = \text{voltage} \times \text{current} \times \text{time}$$

$$P = VI \qquad \text{power} = \text{voltage} \times \text{current}$$

$$\frac{1}{R_p} = \frac{1}{R_1} + \frac{1}{R_2} \qquad \frac{1}{\text{parallel resistance}} = \frac{1}{\text{resistance 1}} + \frac{1}{\text{resistance 2}}$$

$$V_p I_p = V_s I_s \qquad \text{voltage} \times \text{current in primary} = \text{voltage} \times \text{current in secondary}$$

Topic 5
Atomic Physics

Example of α–decay: $^{226}_{88}\text{Ra} \rightarrow ^{222}_{86}\text{Rn} + ^{4}_{2}\text{He}$

Example of β–decay: $^{14}_{6}\text{C} \rightarrow ^{14}_{7}\text{N} + ^{0}_{-1}\text{e}$

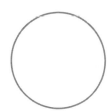

Answers

Topic 1
Try this
1 a) 9 swings b) 5.4 s c) 0.6 s
2 a)

(speed (m/s) vs time (s) graph: line rises from 0 at t=0 to 15 at t=10, then stays flat at 15 up to t=20; y-axis marked 0, 5, 10, 15, 20; x-axis marked 0, 5, 10, 15, 20)

   b) 210 m   c) 10.5 m/s
3 a) 0.8 g/cm^3 b) 2.75 g/cm^3
4 a) 300 N upwards
   b) The rocket accelerates upwards.
5 10 N down
6 a) gravitational potential energy
   b) kinetic energy, strain (elastic) potential energy
   c) 5250 J
7 a) C b) B

Topic 2
Try this
1 The strips will curl with the bronze on the outside.
2 a) 0 °C b) 100 °C c) 47 mm d) 70 °C
3 a) stays the same b) decreases
   c) increases d) stays the same
4 Stage 1 a) increases b) increases c) increases
   Stage 2 a) stays the same b) decreases
   c) decreases
5 a) 60 000 J b) 968 J/kg °C c) 139 g
6 a) conduction b) convection c) radiation

Topic 3
Try this
1 a) 0.75 m b) 0.2 Hz
2 a) towards and away from him
   b) vertically up and down
3 a) B b) E
   c) B = 30°, C = 90°, D = 70.5°, E = 19.5°,
   F = 90°
4 a) real, inverted and magnified
   b) height 3.0 cm (allow 2.8–3.2)
   c) 9.0 cm from the lens (allow 8.8–9.2)
5 a) The two rays diverge and will not meet at
   an image.
   b) i) virtual, erect and magnified
      ii) height 4.5 cm (allow 4.3–4.7)
      iii) 6.0 cm from the lens (allow 5.8–6.2)
6 a) X-rays b) gamma rays c) same
   d) 3.78 × 10^{13} km
7 26 m

Topic 4
Try this
1 a) North b) North
2 a) aluminium b) soft iron c) hard steel
3 a) The field direction alternates.
   b) The force direction alternates.
   c) The cone vibrates at frequency of a.c. signal.
4 a) Charges like those on plastic move to far
   side of stream, unlike charges move to side
   close to plastic.
   b) Like charges: weak repulsive force, unlike
   charges: stronger attractive force
   c) It is deflected towards plastic.
5

Measurement to be taken	AB	CD	EF
Emf of battery	voltmeter	nothing	nothing
Current through R when connected to battery	nothing	nothing	ammeter
P.d. across R when connected to battery	nothing	voltmeter	wire

6 50.3 °C
7 a) 4.4 V b) 0.4 V c) 1.02 V
8 a) 3 A b) 2 A c) 5 A d) 1.2 Ω
9 a) 6 V (or just below 6 V)
   b) 0 V (or just above 0 V)
10

A	B	C	D	E
0	0	0	1	0
0	1	0	0	1
1	0	0	0	1
1	1	1	0	0

11 4000 turns

Topic 5
Try this
1

Isotope	Number of protons	Number of neutrons	Number of electrons
$^{88}_{38}$Sr	38	50	38
$^{90}_{38}$Sr	38	52	38

2

Sample	CR (none)	CR (card)	CR (Al)	CR (Pb)	α	β	γ
1	6000	1000	1000	20	✔		✔
2	3000	3000	20	20		✔	

3 a) deflected downwards b) undeflected
4 $^{238}_{92}$U → $^{234}_{90}$Th + $^4_2$He + γ

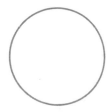

Index

Note: page numbers in *italics* refer to definitions of terms.